Weather and Society

Weather and Society

Toward Integrated Approaches

Eve Gruntfest

Registered Offices
John Wiley & Sons, Inc., 111 River Street, Hoboken, NJ 07030, USA
John Wiley & Sons Ltd, The Atrium, Southern Gate, Chichester, West Sussex, PO19 8SQ, UK

Editorial Office
9600 Garsington Road, Oxford, OX4 2DQ, UK

For details of our global editorial offices, customer services, and more information about Wiley products visit us at www.wiley.com.

Wiley also publishes its books in a variety of electronic formats and by print-on-demand. Some content that appears in standard print versions of this book may not be available in other formats.

Library of Congress Cataloging-in-Publication data applied for

ISBN: 9780470669846

Cover Design: Wiley
Cover Image: © Dreef/Gettyimages

Set in 10/12pt Warnock by SPi Global, Pondicherry, India

Printed in Singapore by C.O.S. Printers Pte Ltd

10 9 8 7 6 5 4 3 2 1

Contents

Acknowledgments

This book began as an effort to support and amplify two ambitious efforts that were passionately dedicated to changing the way that the social sciences and atmospheric sciences collaborate to improve weather forecasting effectiveness, reduce the number of lives lost to severe weather, and to develop a cadre of hybrid socio-meteorologists. The two main efforts toward these goals were the WAS*IS movement (Weather and Society * Integrated Studies) and the SSWIM or Social Science Woven into Meteorology program. When funding for both of these programs ended and I faced the prospect that this textbook would need to be historical and not totally optimistic, progress on this book slowed. The book retains a cautiously optimistic tone even though a close examination in 2017 reveals that atmospheric scientists, without a clear understanding of or appreciation for social science, are doing most of the talking at the intersection of weather and society and there are only a handful of examples of woven collaborations between atmospheric scientists and social scientists.

My graduate school mentor and advisor, Dr. Gilbert F. White, inspired me. Gilbert's brilliance had powerful policy implications for natural hazard mitigation at the local, state, federal, and international levels, and his tireless persistent dedication to bringing all stakeholders, researchers, and students together for decision making kept me focused on getting this book finished. I began my career studying the behavior of people during the 1976 Big Thompson flood in Colorado (Gruntfest, 1977) and the flash flood research led to a broader passion for understanding flash floods and warnings nationally and internationally (Gruntfest and Handmer, 2001). Forty years later, I am still highly motivated and engaged with ways to improve flash flood warnings.

Completing this book required lots of help. I appreciate the feedback and support throughout my career and in the creation of this book of Sandrine Anquetin, Rachael Ballard, Chip Benight, Julia Becker, Eric Beteille, Enrica Caporali, Francesca Carli, Kim Carsell, Julie Demuth, Gina Eosco, Micky Glantz, Neil Gordon, John Handmer, Mary Hayden, Bob Henson, Stephanie Hoekstra, Bill Hooke, David Johnston, Ilan Kelman, Matt Kelsch, Kim Klockow, Emily Laidlaw, Linda Layne, Heather Lazrus, Cedar League, Diana Liverman, Celine Lutoff, Brenda Mackie, Vinodhini Mathiyalagan, Beth Mitchneck, Burrell Montz, Rebecca Morss, Ellen Nelson, Amy Nichols, Lori Peek, Brenda Philips, Andrea Ray, Valerie Ritterbusch, Paola Salio, Celeste Saulo, Lucy Sayer, Fiona Seymour, Andrea Schumacher, Russ Schumacher, Hatim Sharif, Marshall Shepherd, Bill Siembieda, Priya Subbrayal, Bella Talbot and Molly Wingate. My dearest friends, Sarah Christensen, Fanette Pollack, Andrea Herrera, Karen Breunig Hine, Claire Sheridan, and Marc Weber, relentlessly encouraged me to finish this project.

The book is dedicated to the formal and informal WAS*ISers who continue to attempt to weave social science into the fabric of the weather enterprise. I hope this book provides a jumping off point for additions and modifications by new brave generation of scholars, practitioners and charismatic policy entrepreneurs including Simone Balog, Kenny Blumenfeld, Dereka Carroll-Smith, Julia Chasco, Renee Curry, Greg Dobson, Tanja Fransen, Jen Henderson, Eric Holthaus, Rebecca Jennings, Aisha Owusu, Amber Silver Nelson, Daniel Nietfeld, Justin Nobel, Vahid Rahmani, Isabelle Ruin, Jason Samenow, Jen Spinney, Galetia Terti, Chris Uejio, and Charlie Woodrum. These scholars and others are tirelessly leading efforts to integrate social science into atmospheric science, despite limited formal credit for this work from their employers or advisors.

References

Gruntfest, E (1977) *What People Did During the Big Thompson Flood* Working Paper 32 Institute of Behavioral Science University of Colorado http://www.colorado.edu/hazards/publications/wp/wp32.pdf

Gruntfest, E and J Handmer (eds) (2001) *Coping With Flash Floods* Kluwer

Preface

This textbook is the culmination of four decades of collaborations to meet the challenges of natural hazard mitigation. My own contributions came from the social science and societal impacts perspectives. I hope the publication of this book formally marks the end of the era of social science as merely an "add-on" to meteorology.

Until 2005, my expectations for genuine integration of social science into meteorological research and practice were low. I was content to know that at least the words "social science" or "societal impacts" were being considered. I was pleased to be invited to workshops, conferences, short courses, and meetings to be a social conscience. When the presentations went over well, I would be invited back for a similar presentation the following year. Meteorologists would agree that someone should do the social science related to meteorology but no one in particular was volunteering.

I gave dozens of invited talks all over the world. The talks were usually scheduled at the end of a session or a course. While societal impacts and the "so what" part of forecasting and new technologies were acknowledged, the commitment to addressing more than the physical science and engineering was superficial. There was no funding or actual integration of social scientists on research teams. My presentations were not an integrated part of larger research and practice for the physical scientists and engineers in the classes. I always happily fulfilled my role as an add-on social scientist to whatever was the main course, workshop, or conference topic. An egregious example was when I was invited to Germany to provide societal impacts dimensions to the 10-year Mescoscale Alpine Programme (http://www.map.meteoswiss.ch/). Only after a decade of research did the team invite in a few social scientists to provide the "so what" part of the massive experiment. Looking back on my career, preparing to retire from my position as a geography professor, I recognized that, as an add-on social scientist, my work had not made changes in how atmosphere science was being conducted.

Now my expectations for the integration of atmospheric science and the social sciences are higher. The roots of WAS *IS and this textbook can be traced to visiting scientist opportunities I had to work with atmospheric science teams at the National Center for Atmospheric Research (NCAR) and at the Cooperative Institute for Research in the Atmosphere at Colorado State University in Fort Collins, Colorado in the 1990s and early 2000s. I picked up some of their language and concerns of atmospheric scientists that led to grant writing collaborations. In 2012 and 2013, I was invited to serve as a program officer at the National Science Foundation in the Atmospheric and Geospace Division of the Geosciences Directorate. That appointment is one demonstrable piece of evidence among many that there has been movement for the atmospheric sciences to formally appreciate what a social scientist can offer.

So it is with this renewed encouragement that this book lays some groundwork for further integration of social science with atmospheric science. This is just a beginning. Once these two kinds of scientists learn to truly work together, then the real business can begin of more accurately and effectively explaining weather events so that people can respond appropriately and in time.

1

The Need for Integrated Approaches to Weather and Society

1.1 Rationale for This Book

What do you do when you see a weather alert—SEVERE WEATHER EXPECTED? If you are a driver or a passenger in a car, do you pay any attention? Do you change your behavior or do you keep moving forward without any change in your plans? Does your answer depend on where you are heading or how pressing your obligations are? If the alert specifically mentioned icy conditions ahead or wet roadways, would that information be more useful? Would you seek additional weather information? If so, where would you get that information?

Every day and everywhere, people talk about and deal with the weather. They consider the forecasts and the potential weather impacts. The discussions and actions are often place-specific and revolve around what is normal or abnormal for a particular place or region. Conversations about weather range from appreciation of the mild conditions that foster the enjoyment of outside activities to awe and sadness about catastrophic impacts of tornadoes, hurricanes, droughts, and floods. But the question of how to best educate and inform people about weather impacts remains. Some recent weather books for a more general audience that take on this challenge include *The Weather Machine* by Andrew Blum (2017), *Thunder and Lightning: Weather Past, Present, Future* by Lauren Redniss (2015), *The Weather of the Future Heat Waves, Extreme Storms, and Other Scenes from a Climate-Changed Planet* by Dr. Heidi Cullen (2010), and *Weather on the Air: A History of Broadcast Meteorology* by Robert Henson (2010b).

Weather and Society: Toward Integrated Approaches, First Edition. Eve Gruntfest.
© 2018 John Wiley & Sons Ltd. Published 2018 by John Wiley & Sons Ltd.

"Extreme weather is costly. From 2008 to 2013, alone, the price tag of extreme weather events in the U.S was $309 billion. These costs are soaring even as forecasts improve" (Samenow, 2015a,b). Perhaps the most impressive gains in accurately predicting severe weather have been in hurricane forecasting. In the 1980s when the National Hurricane Center tried to predict where a hurricane would hit three days in advance of landfall, it missed by an average of 350 miles. If Hurricane Isaac, which made its unpredictable path through the Gulf of Mexico in August 2012, had occurred in the late 1980s, the center might have projected landfall anywhere from Houston to Tallahassee, canceling thousands of business deals, flights and picnics in between— and damaging the reputation of the National Hurricane Center when the hurricane zeroed in hundreds of miles away. By 2012 the average miss was only about 100 miles (Silver, 2012).

However, even with more accurate predictions, there are severe events that seem to surprise people. More than two feet of rain caused extreme flooding of Baton Rouge, LA in August 2016. The flood killed 13 people, displaced tens of thousands of others, caused an estimated $8.7 billion in damage and destroyed 60,000 houses. More than 73,000 households across 20 parishes were approved for Federal Emergency Management Agency aid for the flooding that is considered to have 1000-year recurrence interval. Linking this extreme event and others to human caused climate change is a lively discussion (Ball, 2016; Mooney, 2016). Should these extreme floods have "names" to help forecasters inform the public about their unusual severity (Shepherd, 2016a; Schroeder *et al.*, 2016)?

Some people have an intense interest in weather details. People who are planning an outside wedding on a particular June day, or who are planning a winter vacation to Florida, or, who are farmers worried how "hard" a predicted freeze will be and its potential impact on their orange groves are particularly weather-aware. Many, but not all, people have a serious, deep interest in weather and meteorological phenomena and they pay close attention to forecasts, warnings, and their own experience.

People use official forecasts from meteorologists, observations from weather stations, and many personal idiosyncratic ways to predict and understand the weather and weather forecasts. Many people rely on environmental cues to tell them whether a storm is coming or if the rainstorm will produce flooding. Some rely on sophisticated scientific and technological tools. Others rely on a combination of sources including low and high technology data.

Some people have little confidence in official forecasts, preferring to "look out the window" or rely on folk methods, "gut readings," or previous experience. Increasingly, more people look at their phones to see radar images of storms moving across town or to listen to the latest official warning from a trusted television meteorologist or the National Weather Service (NWS). Some people consider the continuum of weather from bad to good. What is considered "good" weather? The characterization is related to season and locale. The improvements in forecasting have changed how and when people respond to a major blizzard in 2016 compared to people in 1888.

Research Spotlight Box: Forecasts are Improving Dramatically Impacts of Blizzards in the Nineteenth Century Were Devastating Compared with Impacts in 2016

Imagine that you were a farmer in the Northern Great Plains of the United States on January 12, 1888. You were out in the fields feeding your cattle and then you were getting your children ready for school. The weather was mild. In the afternoon, as school was letting out, the temperature dropped 30 degrees and a major blizzard caused blinding snow and cold winds. More than 100 children died on their way home because they were unprepared for the severe snowstorm that occurred (Laskin, 2004). There were no forecasts of that storm that has been named the "Children's Blizzard." More than a total of 500 people were killed as a result of that "surprise" storm. There were no "means of monitoring the upper atmosphere, no satellites, no radar, no wireless communications with ships at sea, and no computer models for forecasting the weather" (Moran, 2012:2).

Flash forward to 2017. Forecasts of any major blizzard are made days in advance. As a storm gets closer the forecasts become more precise with probabilities of certain amounts of snow or high wind speeds. Satellites, that orbit the earth continuously, monitor a storm's developing movements, radar locates snow bands sweeping onshore and observational data from the surface and upper atmosphere feed into sophisticated numerical weather forecast models running on supercomputers. Winter storm watches and warnings are issued by the National Weather Service (NWS). Some schools announce snow days before the first snowflake falls. People go to supermarkets to stock up on bread, water and batteries. National Oceanic and Atmospheric

Administration's (NOAA's) *four-day* predictions for hurricane track have become as reliable as our two-day predictions were prior to 1995. In 2015, *five-day* temperature forecasts have the same level of accuracy that *three-day* forecasts had 20 years ago. NWS tornado warning lead times have more than doubled over the past two decades, to an average of 13 minutes (Sullivan, 2013:6).

Unlike 100 years ago, school superintendents use many resources, including local social networks, official government websites, private meteorologists, webcams and other resources to help support their decisions of whether to keep schools open or to close early because of existing or predicted "severe" weather. Recent research efforts summarize the various information sources leaders of school districts use. Their decisions have enormous consequences for public safety (Balog, 2013; Call and Coleman, 2012; Hoekstra, 2012; Montz *et al.*, 2014).

In 2017, many atmospheric scientists with classical training in meteorology recognize that improved forecasting is not the most serious challenge to reducing losses from severe weather, the biggest challenges arise from predicting human behavior in response to these weather events or forecasts. These scientists appreciate that the severity of weather's impact has at least as much to do with the capability of the population at-risk to reduce its own vulnerability as it does with the strength of winds, the height of a storm surge, the high temperatures, the depth of flood waters, or the strength of a tornado. A perfect flash flood warning will not affect the behavior of everyone who hears it. People are aware of the risks of driving across flooded roads and know there are warnings in effect but sometimes they HAVE to get somewhere—to work or to pick up the kids. Or, maybe they want to test whether their truck can successfully forge flooded roads (League, 2009; Ruin, 2008). Social scientists can help assure that the forecasts reach the vulnerable people and increase the likelihood that vulnerable people take the appropriate actions in the time they have before the severe weather impact. "An excellent weather forecast, if not properly communicated and acted upon, is of practically no value" (Samenow, 2015a).

This book is the first textbook to share an understanding of how social scientists are working on weather-related problems. It is not exhaustive or comprehensive in its coverage of all the studies at the intersection of weather and society, but it is representative of the range of scholarship that has been completed and is underway. Many new people are getting involved in this work, so there are simply too many research and operations projects to cover in one book. Also, every day presents us with more flash floods, droughts, tornadoes, or snowstorms to learn about and from.

This book provides an overview and is meant to serve as an introduction to the emerging field of socio-meteorology. It highlights historic and contemporary collaborative efforts between social scientists and meteorologists at the intersection of weather and society. This book explores the leading edge of weather research that addresses the impacts of forecasts and warnings. It is a 2018 snapshot of a quickly changing landscape where the characters are changing from primarily governmental and academic partners to a dynamic mixture of governmental agencies, academics, and private companies. This book addresses the numerous calls for widening the community at the intersection of weather and society. It provides a first exposure that can broaden academic programs in meteorology, hydrology, environmental studies, geography, natural hazards, anthropology, communication, economics, and sociology across the globe where there is a growing appreciation of more interdisciplinary and multi-disciplinary approaches to answering questions about how to reduce the negative impacts of weather. It offers an overview of the growing body of literature that tries to solve problems related to weather and society.

Weather and Society: Toward Integrated Approaches is written for anyone who wants to learn about how to think about integrating social science and atmospheric science. This book starts to address the needs of a growing community of people who want to learn about how the social sciences can contribute to solving problems at the intersection of weather and society. Tackling these complex problems, including reducing losses from weather events, calls for interdisciplinary cooperation and multi-disciplinary approaches. In 2014, the American Meteorological Society (AMS) Board on Societal Impacts adopted a professional guidance statement aimed at strengthening social sciences in the weather-climate enterprise (http://www2.ametsoc.org/stac/index.cfm/boards/board-on-societal-impacts/).

Providing the groundwork for new ways for meteorologists and others to approach their own disciplinary and interdisciplinary challenges, this textbook complements courses offered by various professional societies and professional associations including the National Hydrologic Warning Council (www.hydrologicwarning.org), the American Meteorological Society (www.ametsoc.org), the Association of State Floodplain Managers (www.floods.org), the International Association of Emergency Managers (www.iaem.com), COMET (www.comet.ucar.edu), and other professional groups that have online and in-person certification courses. All of these groups are seeking ways to incorporate lessons from social science to improve warnings, interagency communication, interdisciplinary partnerships, public and private relationships, and others.

Meteorology textbooks cover a wide range of atmospheric physics principles and applications. They do not emphasize societal impacts of weather or results of social science research studies related to weather. They often use case studies of extreme historical events with an emphasis on the meteorological characteristics such as wind speeds, amount of precipitation, or hail size, but atmospheric science textbooks do not include chapters on how people are affected by the weather or how research by social scientists can reduce vulnerability to severe weather or increase understanding of weather forecasts.

What people do when facing severe weather or when warnings are issued is a central topic of this book. This book provides various considerations of how the NWS is changing the ways its forecasts and warning products are issued to respond to the weather information needs of decision makers including school administrators, emergency managers, ranchers, transportation departments, and others. The discussion of "warnings" in traditional meteorology textbooks covers the products issued by the NWS and mentions briefly the recommended public behavior that the warnings should influence, but there is no impacts section, social science, or societal impacts section (Moran, 2012).

Weather and Society: Toward Integrated Approaches starts with the ways social scientists and others are learning about how people behave in severe weather. This book takes a 30,000-foot or 10,000-foot view of the issues at the intersection of weather and society. It provides a big picture of the societal impacts of weather and how social scientists can and do collaborate with meteorologists to address weather challenges that require combined atmospheric and societal approaches. It is written for an audience that recognizes the importance of bringing the applied aspects, or the "so what" factors, together with the more physical science of forecasts.

This book builds on the growing recognition from academia, government, the private sector, and the non-governmental sector that when meteorologists and social scientists work together, they provide results and products that are greater than the sum of their individual parts. Many early career physical and social scientists and engineers interested in weather seek to learn about concepts, tools, questions, and policies related to more than one discipline, but most academic departments, especially in meteorology and hydrology, are too narrow to allow much leeway for electives outside of their narrowly defined discipline.

Since 2000 the integrated atmospheric-social science community has relied on informal social media networks that have risen in popularity and proved their usefulness. As of 2018, these social networks include Facebook, Twitter, NWS chats, and blogs as the main sources of data and information, since more standard, classical sources tend to remain quite narrow in their uni-dimensional or uni-disciplinary approaches. It is time for a more formal and organized approach.

This book includes new perspectives from a wide variety of specialties that are taking into account the societal impacts of weather and developing new ways of thinking about forecast verification. It means stepping into new points of view and being open minded to new ways of seeing weather and its impacts. It presents ideas that are being developed by a community of scholars and practitioners dedicated to changing the stove-piped culture of current scientific disciplines. This community is devoted to new problem-solving approaches and is willing to take the time necessary to learn the languages and perspectives of people from different backgrounds and disciplines. This book uses case studies to highlight the complexity and multi-dimensional aspects of some of the most pressing problems at the intersection of weather and society. The book shows how different social scientists have framed and represented their integrated work. Using many figures from recent publications and presentations provides the "look and feel" of how integrated work is conducted and reported.

Weather and Society: Toward Integrated Approaches has seven main goals:

1) To create an environment where scientists and teachers can develop materials for stand-alone meteorology or weather-society courses or to supplement current materials with a social science dimension;
2) To provide the groundwork for conducting interdisciplinary work by learning new strategies and addressing typical challenges;
3) To expose the readers to various social sciences, the methods they use, and the ways they share their data;

4) To identify research, application, and educational opportunities for integrated weather-society work;

5) To review the major institutional efforts to bring social science and social scientists into the research and practice of meteorologists and hydrologists in sustainable ways;

6) To show central topics for the new hybrid socio-meteorologists; and,

7) To provide a summary of key challenges and directions for work in the near and distant future.

This book aims to jumpstart the dialogue between all the partners. The vision for this textbook is that "it has something for everyone but it is not everything for anybody." It is a smorgasbord of people and their activities at the intersection of weather and society. This book shows students, faculty members, forecasters, emergency managers, broadcast meteorologists, and many others who never have seriously thought about their studies or their work in this new context that there are enormous challenges as well as career opportunities at the intersection of weather and society.

Weather affects so many aspects of daily life and researchers from many disciplines research weather impacts. Weather and crime (Ranson, 2013); weather and art (e.g., Thornes, 2008); weather and tourism (e.g., Denstadli *et al.*, 2011; Sabir *et al.*, 2014; Martín, 2005; Jeuring and Becken, 2013); weather and election day turnout (e.g., Persson *et al.*, 2014); weather and fashion (Hershey, 2015); weather and traffic accidents (e.g., Hranac *et al.*, 2006; Dell'Acqua *et al.*, 2012; Strong *et al.*, 2010; Cai *et al.*, 2013); weather and the number of broken bones a hospital can expect to treat (e.g., Murray *et al.*, 2011); weather and its effects on home health care providers (Joseph *et al.*, 2012; Skinner *et al.*, 2009); what kind of weather makes people sad (e.g., Huibers *et al.*, 2010); and even how moods are related to stock market activity (Früwirth and Sögner, 2015)—these are research topics at the intersection of weather and society.

1.2   The Audience for This Book

Many meteorology, hydrology, and ecology students are asking about the social aspects of weather. They observe how professional societies like the National Weather Association (NWA) organize sessions to reach out to the victims of recent weather disasters such as the 2011 town hall session dedicated to learning from the experience of the Tuscaloosa tornado victims. Other students become interested in the social aspects of meteorology because of some disaster that hit close to their home or caught their attention and concern through news reports (www.nwa.org).

More and more graduate students, potential graduate students, NWS employees, hydrologists, public safety officials, and others recognize that some understanding of social science and societal impacts can enhance their work as practitioners, researchers, or students. They also realize that it is frustrating and difficult to find reference materials or courses to meet their need within the traditional university departments of meteorology, atmospheric science, hydrology, or even physical geography. This book's intended audience is anyone who wants to learn about the intersection of atmospheric science and the social sciences. The book aims to reach out to students who represent the next generations of scientists and practitioners with the intent to provide a platform for new ways of approaching pressing problems at the intersection of weather and society.

The topic of the societal impacts of weather is becoming more visible. Each year at national meteorology conferences, including the American Meteorological Society and the National Weather Association, there are more professionals who are presenting research findings related to improving communication to the public. They recognize that the societal impacts of weather is an important and upcoming field especially as the set of normal conditions is replaced by new "normals" in terms of extreme weather as the impacts of climate change intensify and social media offers many options new real-time communication of weather conditions and forecasts.

This book shows examples of recent work, but unfortunately as of 2017 there is no academic program specifically focusing on the societal impacts of weather. Even without official programs and sanctions, the community of practitioners and researchers who work in integrated teams or who appreciate the value of interdisciplinary collaboration at the intersection of weather and society is getting larger and more diverse. Even though the field is in its infancy relative to the body of atmospheric science literatures as a whole, the social science studies available are too numerous to describe all of them. For example, as of June 2015, a Googlescholar search for articles related 2005's Hurricane Katrina and social science, shows 50,400 links overall (accessed June 10, 2015).

In this rapidly developing field, every day brings new large and small research findings, changes in forecasting operations, and reports written and published by students and researchers from agencies and universities that increase understanding of how weather information can be most effectively packaged and communicated for all of us who make weather-sensitive decisions.

This book is not a roadmap with a clear set of boxes to check off showing how to do integrated weather and society work. The innovative ways forward will be built with the ideas, practices, and projects that are being imagined by the people and institutions that embrace integrated approaches. When funding agencies, university departments, weather companies, and agencies approach problem-solving with new principals that incorporate qualitative and quantitative methods and elements from meteorology and social science disciplines, new effective approaches will go far in solving the most difficult problems facing the weather community. This will take patience, capacity building (especially bringing in new people with social science backgrounds) cooperation, funding, and a genuine appreciation that when teams are formed from the start with equal numbers of physical and social scientists, we will achieve the goals of reducing losses from severe weather.

Even with some attention to the international work that is underway, this book takes a U.S.-centric approach that relies on examples from the NWS and other U.S. agencies and research institutions. Consider this book a starting point for the next book, which will have a broader global perspective and perspectives of many other researchers and practitioners working on problems of weather and society.

1.3 Defining Weather and Society: Integrated Approaches

Figure 1.1 represents the numerous sets of partners engaged in weather and society work. There is no starting point, and there is no top or bottom. The partners are all engaged in their work without any hierarchy of purpose.

In 2017, many people in the meteorology community still place the hope for reducing losses from extreme weather events on the promise of new expensive technologies that are expected to make "all the difference" in improving weather forecasts. Since the 1960s, the new

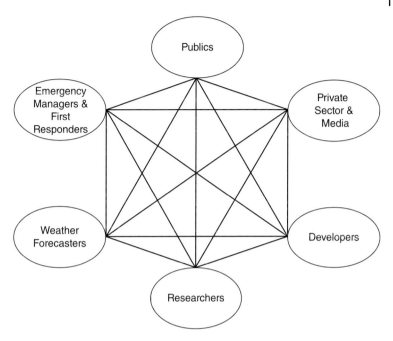

Figure 1.1 The major groups that are concerned about weather and society. It provides a basic diagram of the weather community. It is meant to include everyone who is concerned about the weather, from those who design tools to forecast the weather, to those who do the forecasting, to those whose daily professional and personal lives depend on the weather forecasts and weather impacts. Partners in integrated weather and society work are all connected and have different spatial and temporal information needs.

technology could be automated systems, new more powerful or adaptive radars, new computer programs, or new structural control works. Often stakeholders are told that once the new software was in place, or the new radar, or the new gauge network, that progress in forecasting and mitigating losses would emerge. Despite the implementation of many new technologies, we still have many weather-related losses. Technology alone is not the silver bullet.

Buying the technology is the easy part. Effective warning systems must have working detection and response components. Until we can motivate more people to take appropriate actions in ways and in time to reduce their vulnerability, major benefits of the new technologies go unrealized. There is a growing recognition that progress

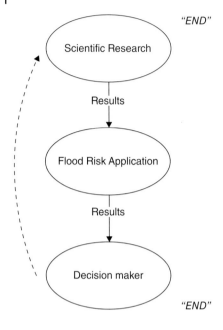

"END"

Figure 1.2 Scientist's typical view of research and development to produce useful information for society—top-down "end-to-end" research illustrated for the case of flood risk management. The connection from decision maker to research (shown by dashed line) is mentioned in some implementations of end-to-end research, but in others, it is left implied or assumed. *Source:* Morss *et al.*, 2005:1598. Reproduced with permission from the American Meteorological Society.

"END"

requires the end-to-end-to-end approach, engaging all partners from individuals, local officials, first responders, researchers, and others. A top-down approach has proven unable to effect the needed changes, across the weather enterprise, to reduce losses to the extent possible.

Creating the needed integrated approach to weather and society requires collaboration between physical and social scientists. Dr. Rebecca Morss and her colleagues have published several articles detailing the necessity to consider end-to-end approaches. Their work aims to widen the perspective of meteorologists beyond the traditional approach of doing scientific research disconnected from the stakeholders or decision makers who might be able to use or critique the research. This traditional model (Figure 1.2) shows how scientific research is done and the results are fed "down" to the decision maker. This top-down model is in contrast to working in a process that can be called end-to-end-to-end, where the scientists and decision makers interact as part of the scientific process (Figure 1.3). The two ENDs in Figure 1.3 represent the two ENDs in top-down research; end-to-end-to-end research signifies iteration between these two ends (Morss *et al.*, 2005:1599).

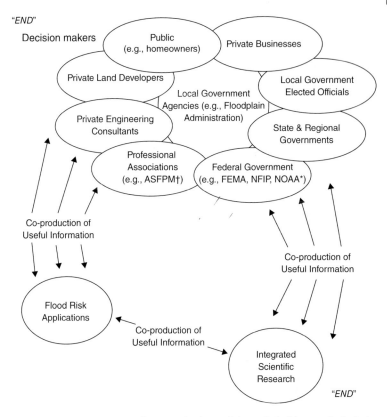

Figure 1.3 A more integrated approach where all the stakeholders are included in the process and the process has no top or bottom. This iterative process is slowly taking hold as meteorologists recognize the value of developing new tools or conducting research with stakeholder preferences or concerns in mind. *Source:* Morss *et al.*, 2005:1999. Reproduced with permission from the American Meteorological Society.

Some physical scientists underestimate social science. Generally, they do not understand the methods of social science or their importance. Physical scientists are not trained in how to write thoughtful questionnaires or survey questions. They do not realize that focus groups require careful sample selection and serious attention to the phrasing of questions to be sure that the issues that need to be addressed are being communicated effectively. A meteorologist needs to find a trained social scientist at a local university when he or she wants to find out how the public is using a forecasting product, like a watch or a warning, or whether emergency managers are getting the forecasting information they need to make timely decisions. Dr. Morss and her colleagues illustrate the end-to-end-to-end (not top down) iterative process in a study of floodplain management options.

Figure 1.3 provides a revised view of research to produce information that is useful in one or more specific societal applications: "end-to-end-to-end" research, illustrated for the case of flood-risk (specifically floodplain) management with diverse, interconnected decision makers. The end-to-end-to-end approach explicitly recognizes the importance of multidirectional communication; sustained interactions among researchers, application developers, and multiple decision makers. This model highlights that it takes multiple iterations around the loop to coproduce knowledge and tools. Integrated scientific research includes disciplinary and interdisciplinary work in statistics, climatology, meteorology, hydrology, engineering, geography, and the social sciences and humanities.

When physical scientists have moved forward without the aid of social scientists, they often find that the data they collect do not answer the most important questions in the eyes of the decision makers. Physical scientists posing as social scientists, constructing their own surveys often very quickly, often think their findings represent a point of view when in truth the respondents to the survey do not understand what they were being asked and the findings are not useful. Those in the field are increasingly recognizing that progress requires the end-to-end-to-end approach engaging all partners from individuals, local officials, first responders, researchers, and others.

The 2010 Board on Atmospheric Science and Climate's (BASC) National Research Council (NRC) Report *When Weather Matters* was the first formal publication to call for equal footing between physical and social scientists. The report's recommendations are evidence of a growing understanding of how social science can help address the most pressing weather, climate and society challenges. "The weather community and social scientists should create partnerships to develop a core interdisciplinary capacity for weather-society research and transitioning research to operations, starting with three priority areas: estimating the societal and economic value of weather information; understanding the interpretation and use of weather information by various audiences; and applying this knowledge to improve communication, use, and value" (BASC, 2010:2-3).

Social science contributes understanding of how the social and economic value of weather and climate information is measured and used by various audiences to make decisions and reduce vulnerability, and it contributes to understanding how specific populations perceive and adapt to weather and climate (BASC, 2010). As of 2017, the NOAA and NWS Strategic Plans include calls for greater integration of social science research and results. In the United States and elsewhere, numerous federal and state agencies and university groups are promoting interdisciplinary approaches at the interface of society, weather, and climate. *When Weather Matters* called for integrated research and programs that address meteorology and social science. "The BASC committee's vision is that by ~2025, a core group of social scientists and meteorologists will have formed a strong, mutually beneficial partnership in which multiple areas of science work together to ensure that weather research and forecasting meet societal needs. The knowledge and expertise needed to address critical problems at the weather–society interface efficiently and reliably will be readily available, and it will be applied regularly to address research questions of interest to both social scientists and meteorologists and to enhance weather Research-to-Operations and operations themselves" (BASC, 2010:2).

In 2017, Argentina has been working to develop new weather warnings called ALERT.AR that start with what first responders and emergency managers are already using. The Argentine Weather Service wants to improve warnings, and they recognize that improvements are in the eyes of the stakeholders—not in the eyes of the people who invent the forecasts (Saulo, 2015; Chasco, 2016). This promising shift of focus is the sort of first step toward understanding that warnings work better when the people who use them understand them better.

1.4 What Social Sciences Have in Common with Each Other and with Atmospheric Science?

The social sciences are a group of academic disciplines that study the many different ways that people organize and live (Figure 1.4). Like physical scientists, social scientists rely on the scientific method. As all scientists do, they use observation, develop and apply theories,

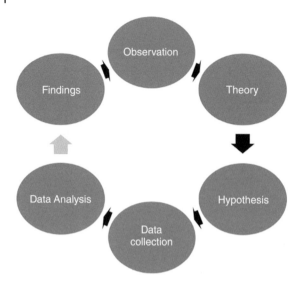

Figure 1.4 Like meteorology, the various social sciences rely on the scientific method and use many different methods for their research.

make hypotheses, collect data to test their hypotheses, analyze their data, develop findings based on their data analysis, and then draw conclusions.

Several of the social sciences, discussed in Chapter 2, work on problems at the intersection of weather and society. There is wide variety of weather-related questions that social scientists tackle. A small sample of the types of questions being addressed is listed here:

1) Where do professional decision makers and unsophisticated weather information consumers (school teachers, parents, taxi drivers, business owners, and others) get their weather information and how much confidence do they place in the official forecasts? How often do they confirm warnings by talking with friends or family or turning to television or the internet?
2) How much time is needed for cities and families to evacuate prior to a hurricane? Does forecasting lead time accurately convey what the meteorologists know about when and where impacts will occur?
3) How much value do weather forecasts provide to the public relative to the amount of funding that is allocated to the agencies that develop and create the forecasts?

The social sciences are sometimes incorrectly grouped together by non-social scientists as having a unified worldview. Many atmospheric scientists see all social sciences (and policy studies) as an undifferentiated whole entity. When recommendations are made to "incorporate social science" into meteorology, rarely is there clear idea of what that work would look like. Because they tackle the questions listed above, the social sciences, like physical sciences, are scientific and diverse.

Until recently, the only differentiated social science that was incorporated in meteorology, if one was offered at all, was economics. Economic studies can assess the "value" of a particular weather product. Atmospheric scientists have worked with economists to place an economic "dollar value" on forecasts and lead time. Without a detailed collaboration between the meteorologists and other social scientists, research results from the perspective of one social science might be the only social science information pursued. For example, even when considering "value" of a weather product, economic value is only one way of measuring importance. There are also cultural, social, historic, and intrinsic values to take into account. There also are direct and indirect values. When someone's home is destroyed, that loss is fairly easy to place in a dollar amount. However, when the university or a hospital is destroyed, how are the losses in terms of jobs calculated and what are the multiplier effects when many city residents are forced to relocate or shop or work in other cities when their own facilities are destroyed by a weather event? These costs are considered "indirect" costs and they can be very significant, long lasting and difficult to pinpoint, and economics alone cannot do it.

A thoughtful look reveals that each social science has a rich history of theory, methods, and applications with outstanding experts in each field and a depth and diversity often unknown to those outside of the discipline even within other social sciences. Examples of interdisciplinary collaborations are expanding with varying levels of integration and success. Social sciences are varied and study a wide range of aspects of human society. Social scientists deal with people as individuals, families, and institutions and in political contexts. Social sciences and policy are intimately related, and they focus on different topics. Just as physical scientists have difficulty collaborating across disciplines or with social scientists, social scientists have similar difficulties collaborating with each other.

Most people who want to work at the intersection of meteorology and social science will collaborate with social scientists from one of the numerous social science disciplines. These disciplines are worth presenting in some detail since few physical scientists have ever been exposed to any of the social sciences; more knowledge can challenge the perspective that all social scientists have one tradition.

This textbook shares some social science studies from a list of select disciplines. Chapter 2 provides case studies of weather/society research from anthropology, communication, economics, geography, psychology, and sociology. The summary of six disciplines is a first step to understanding that there are major differences between the types of problems that various social science disciplines tackle. A disciplinary perspective is useful for two main reasons: 1) it helps to understand the theoretical frameworks and key questions of the discipline and the people who have been trained within that tradition, and 2) most academic institutions, scholars, and practitioners identify within the traditional disciplines rather than from an integrated, problem-solving perspective.

1.5   Social Science Methodologies

Social scientists use many methodologies including qualitative and quantitative approaches. Qualitative research methods provide detailed insights into decision making or behavior. Dillman has written the authoritative text on social science methods. He has updated his book to include internet-based surveys and use of other social media (Dillman, 2007).

1.5.1   Surveys

Social scientists rely on structured, semi-structured and unstructured interviews. Social scientists invest time in developing comprehensive surveys. It is easy for non-social scientists to underestimate how much time it takes to create a thoughtful survey. Survey development requires more than a few minutes writing a set of questions. Each question must be constructed to be sure that it is explicit, not confusing, and gets the type of information the study requires. Survey design is an art in and of itself.

Before conducting the survey, a sample must be carefully selected. Samples can range from complete coverage of the entire population or can be selected subgroups. Surveys can be administered online, in person, over the telephone or through the mail. Recent social science research findings have been obtained though internet surveys (e.g., Morss *et al.*, 2008). Earlier research relied on in-person, phone, or mail-in surveys.

1.5.2   Direct Observations

The ubiquitous cameras are a valuable source of data about weather impacts and how people behave when faced with "bad" weather. Before there were so many cameras at intersections, homes, commercial enterprises, and in other locales, direct observation required time-intensive personal face-to-face interaction.

Cedar League's master's thesis is an example of another type of observation. She did not go into the field. She looked at YouTube videos showing people driving across flooded roads. Her geography research involved studying the videos to find out many behavioral and demographic aspects (e.g., if people were alone, what kind of vehicle they were driving) (League, 2009). She then contacted many of the people who posted YouTube videos and asked them questions about why they were driving across flooded roads and if they knew about risks and warnings. Nearly half of her respondents said that they "needed" to get somewhere and the other half said that they wanted to "see if their vehicle could make it." League's research shows how research using YouTube and other social media sites can help forecasters and emergency managers understand why people behave as they do.

1.5.3   Participatory Action Research

An increasingly used methodology, participatory action research seeks to involve research subjects in solving problems that they identify affecting their communities. One such problem is dealing with disasters (e.g., Mercer *et al.*, 2008; Gaillard, 2010). The words in this kind of research highlight three main elements: 1) the research subjects are full participatory partners in the work of trying to solve a problem, 2) action to solve the problem needs to arise from the work, and 3) original science, that is, research, is still being produced. One of the problems identified for solving is how people deal with disasters. Participatory action research is increasingly being used to determine and apply how vulnerability could be reduced over the long term.

Participatory action research would determine the individual and collective choices that the occupants and the occupants' communities make within the web of local, national, regional, and international influences that created and continue to perpetuate a long-term situation of vulnerability. "Community" at all scales is incorporated, from the occupants' neighbors to the national government and international institutes. Decisions over all time scales are also included, from day-to-day acquisition of food to century-to-century decisions of where to live. Rather than relying on one focus, one discipline, one knowledge base, one group of people, or one technique, a combination and balance is needed. The focus of solving the problem is action, to improve the situation so that people do not experience similar vulnerability or damage in the future. Decisions are made and reported not just by researchers or practitioners, but also by the occupants themselves and their communities.

1.5.4   Focus Groups

A focus group is a guided discussion where a moderator leads a group of participants through a set of questions on a particular topic. Focus groups are often used in the early stages of projects to obtain feedback about users, products, concepts, prototypes, tasks, strategies, and environments. Focus groups can also be used to obtain consensus about specific issues. Focus group moderators generally follow a discussion plan that has the questions, prompts, tasks, and exercises for the group. The success of a focus group is heavily dependent on the skill of the moderator. The moderator must generate interest in the topic, involve all the participants, keep the discussion on track (but also allow for unexpected diversions), keep dominant personalities from overwhelming other participants, and not give away the sponsor's beliefs or expectations (http://www.usabilitybok.org/methods p866). A facilitator leads a guided discussion of 6 to 12 people on a specific topic. A typical focus group normally lasts one or two hours, is normally recorded and a report is produced of the process and results. The focus group may be watched by the "client" or other interested parties. Focus groups provide useful information on how people respond to particular questions or issues, but the short amount of time limits the depth of discussions. Follow-up focus groups or in-depth interviews are useful to get more detail on perceptions or attitudes.

NOAA's Office for Coastal Management DigitalCoast website has many specific recommendations for ways to conduct social science research. For example, they have an "Introduction to Conducting Focus Groups" that is available online and in print for free (https://coast.noaa.gov/digitalcoast/training/focus-groups.html).

In 2003, focus groups proved very useful in the development of public education materials related to reducing the impacts of an outbreak of West Nile Virus in Colorado. (http://bcn.boulder.co.us/basin/watershed/westnile.html). Many people experienced severe symptoms ranging from loss of memory to paralysis, and 63 people died in Colorado. According to the epidemiologists and public health officials, people who were over 50 years old were most vulnerable to the severe symptoms. The state developed a series of public education materials for the "elderly" to encourage them to avoid exposure to the mosquitoes: encouraging them to stay inside at dawn and dusk, encouraging them to use repellent including DEET and wearing long-sleeved shirts and long pants (https://www.colorado.gov/pacific/sites/default/files/DC_CD_Zoo-WNV-infection-prevention-and-control-recommendations.pdf.) However, the focus groups revealed that these materials were ineffective because no one identifies as elderly. One person in the focus group who was 80 years old said that her mother would be considered "elderly." Because effective public education must be aimed at an audience that can personally identify with the campaign, these materials did not do as much good as they could have.

1.6   What Is Not Social Science?

Many people say they are doing social science when they are practicing something else. Finding out what the elements are of a new forecasting product is market research. It is not the same as social science. Social science is not someone writing a survey in five minutes and sending it to her Facebook friends. Some people think incorrectly that social science is inexpensive, quick, and easy and requires no technical training. As some of the case studies will reveal, social scientists are often brought too late into the process of understanding whether or not a new forecasting product is effective. This means that the agency has already developed the new product, for example, a storm-surge mapping tool, and has already heavily invested in its implementation. Bringing social scientists into the process earlier assures that the intended users of the product would find it useful and timely before the agency's time and resources have been squandered.

1.7 Doing Social Science Versus Incorporating Societal Impacts

There is a difference between doing social science and measuring societal impacts. People who are measuring societal impacts of weather and recording damages and deaths may feel they are doing social science, but since they are not conducting any scientific research, they are really just accounting for the weather impacts. Measuring impacts is essential to understanding how weather affects our communities, but it should support, rather than take the place of, social science. Social science research helps clarify important definitions; for example, such research reveals the difference in the meaning of "severe" weather for an emergency manager and for a NWS forecaster.

That the first snow of the winter season, especially if it occurs earlier than usual in the year, is a good example of how studying impacts to support social science research can make forecasts more useful for their users. Research on how people use forecasts shows us that a useful forecast for an early storm will emphasize to drivers to be aware that they might be expecting the roads to be wet when they are icy. The same storm later in the year would not have the same impacts when drivers and pedestrians are more prepared for winter so the forecast doesn't need to mention the ice. Winter snow storms that meteorologically are "nuisances" because they have low accumulations can have serious consequences for the people using the forecast if the three inches occur in a 15-minute period during the morning or evening commute period. New NWS criteria recognize that the impacts of a storm on Washington, DC are more severe than the impacts of the same amount of snow in Duluth, Minnesota. Generally, a Winter Storm Warning is issued if at least 4 inches (10 cm) to 7 inches (18 cm) or more of snow or 3 inches (7.6 cm) or more of snow with a large accumulation of ice is forecast. In the southern United States, where severe winter weather is much less common and any snow is a more significant event, warning criteria are lower, as low as 2 inches (5.1 cm) in the southernmost areas. Seasonality also plays a role in storm impact severity. A warning can also be issued during high impact events of lesser amounts, usually early or very late in the season when trees have leaves and damage can result. While most forecasters and others might assume that if a severe thunderstorm is expected to occur at 1:00 a.m. that the societal impact would be low since most people would be home in bed. However, Amy Nichols found in her interview with a university emergency manager that if he gets a warning for a 1:00 a.m. storm, he knows that his constituents, students at his university, might be planning to walk home from clubs or bars and he would use social media to notify them of the warning. If the storm were expected to occur at 3:00 a.m., he would not need to notify the student because the bars would already have closed at 1 a.m. and he would expect them to be already at home (Nichols, 2012). These are two examples of the value of using social science research to create "impacts-based" forecasts for the NWS.

1.8 Questions for Review and Discussion

1 Would you consider a large tornado or flood a disaster if no humans were killed or buildings destroyed? What about an event with severe environmental effects but no immediate loss of life or property? For example, consider a cut in a barrier island or the destruction of marshes protecting the seashore? How do you compare the impacts of these events with an event where lives are lost? Justify your answer with three main points.

2 Do you consider short fuse events, such as tornadoes, to be more severe than longer fuse events, such as droughts? What about unseasonable events such as an ice storm that occurs earlier than normal in September or a tornado that occurs in January prior to what is considered normal tornado season?

3 Do you agree with the premise of this book that more people need to understand elements of atmospheric science and the social sciences? Defend your answer with two reasons.

4 How do you think population density relates to weather vulnerability? Is vulnerability related solely to the numbers of people or are there characteristics of the population that affect their levels of

vulnerability? Discuss using a case study such as Hurricane Katrina or a more recent event closer to where you live.

5 In the Morss diagram of stakeholders (Figure 1.3), who else do you think should be involved? Are there local policy makers or businesses that would be considered key stakeholders where you live? The "public" consists of many different subpopulations including people who need extra time to evacuate because of age or disability or many large pets, people who do not speak English, nursing homes, and other special facilities and many other subsets. Who can you add? How would you expand the single bubble for the "Public" to make the diagram even more reflective of society? What about utility companies? How might the model change if the hazard was a tornado, winter storm, lightning storm, or hurricane?

6 In the broadest context at the global level, in what ways do meteorological phenomena shape society (societies)? What counts as "normal" weather where you live? What counts as "abnormal" weather? What counts as "severe" weather? Do your definitions match with the formal agency forecasting definitions? You can find the NWS definitions at: http://www.weather.gov/bgm/severedefinitions.

1.9 Using What You've Learned: Homework Assignment From the Chapter

1 Pick one of the research questions posed in the section "What social sciences have in common with each other and with physical sciences?" or choose your own research question. Then choose a social science research method (or methods) you would use to address the question and briefly explain why you would use this method?

2 Where do professional decision makers (such as transportation planners or snow plow drivers) and unsophisticated weather information consumers (school teachers, parents, UBER drivers, business owners, and others) get their weather information and how much confidence do they place in the official forecasts? How

often do they confirm warnings by talking with friends or family or turning to the TV or social media? What apps do you have on your phone that are most useful to your personal decision making related to weather?

References

Ball, J.R. (2016) Louisiana Flood of 2016 made worse by growth-focused policies. *The Times-Picayune*, September 23.

Balog, S. (2013) *Severe Weather Decision-Making: A Study of Headteachers in Wales and Western England.* Unpublished master's thesis, King's College, London.

Blum, A. (2017) *The Weather Machine.* New York: Ecco/Harpercollins.

Board on Atmospheric Science and Climate (BASC), Committee on Progress and Priorities of U.S Weather Research and Research-to-Operations Activities (2010) *When Weather Matters: Science and Service to Meet Critical Societal Needs.* National Research Council. http://www.nap.edu/catalog.php?record_id=12888 (accessed July 23, 2017).

Cai, X., Lu, J.J., Xing, Y., Jiang, C., and Lu, W. (2013) Analyzing driving risks of roadway traffic under adverse weather conditions: In case of rain day. *Procedia - Social and Behavioral Sciences*, 96: 2563–2571.

Call, D., and Coleman, J.S. (2012) The decision process behind inclement-weather school closings: A case-study in Maryland, USA. *Meteorological Applications*, 21:474–480.

Chasco, J. (2016) Alert.AR Ask to Answer Servicio Meteorologicó Nacional. PowerPoint presentation.

Cullen, H. (2010) *The Weather of the Future Heat Waves, Extreme Storms, and Other Scenes from a Climate-Changed Planet.* New York: Harper Collins.

Dell'Acqua, G., De Luca, M., Raffaele, M., and Russo, R. (2012) Freeway crashes in wet weather: the comparative influence of porous and conventional asphalt surfacing. *Procedia Social and Behavioral Sciences*, 54: 618–627.

Denstadli, J.M., Jacobsen, J.Kr.S., and Lohmann, M. (2011) Tourist perceptions of summer weather in Scandinavia *Annals of Tourism Research*, 38(3): 920–940.

Dillman, D.A. (2007) *Mail and Internet Surveys The Tailored Design Method Update with New Internet, Visual, and Mixed-Mode Guide.* New York: Wiley.

Frühwirth, M., and Sögner, L. (2015) Weather and SAD related mood effects on the financial market. *The Quarterly Review of Economics and Finance,* 24 February (online).

Gaillard, J-C. (2010) Participatory mapping for raising disaster risk awareness among the youth. *Journal of Contingencies and Crisis Management,* 18: 175–179.

Henson, R. (2010b) *Weather on the Air: A History of Broadcast Meteorology.* New York: AMS Press.

Hershey, L. (2015) High fashion. High Stakes, 20 February, http://www.weatherandeconomics.com/2015/02/20/fashion-industry/ (accessed July 23, 2017).

Hoekstra, S.H. (2012) *How K-12 school district officials made decisions during 2011 National Weather Service tornado warnings.* Master's thesis, Department of Geography and Environmental Sustainability, University of Oklahoma.

Hranac, R., Sterzin, E., Krechmer, D., Rakha, H., and Farzaneh, M. (2006) *Assessing the impact of weather on intensity Empirical Studies on Traffic Flow in Inclement Weather.* Publication number FHWA-HOP-07-073. U.S Department of Transportation Federal Highway Administration.

Huibers, M.J.H., Esther de Graaf, L., FPML Peeters, F.P.M.L., and Arntz, A. (2010) Does the weather make us sad? Meteorological determinants of mood and depression in the general population. *Psychiatry Research,* 180(2): 3143–3146.

Jeuring, J., and Becken, S. (2013) Tourists and severe weather – An exploration of the role of 'Locus of Responsibility' in protective behaviour decisions. *Tourism Management,* 37: 193–202.

Joseph, G.M., Skinner, M.W., and Yantzi, N.M. (2012) The weather-stains of care: Interpreting the meaning of bad weather for front-line health care workers in rural long-term care. *Social Science and Medicine,* 91: 194–201.

Laskin, D. (2004) *The Children's Blizzard.* New York: Harper Collins.

League, C. (2009) *What Were They Thinking? Using YouTube to Observe Driver Behavior While Crossing Flooded Roads.* Unpublished master's thesis, Applied Geography, University of Colorado at Colorado Springs.

Martín, M.B.G. (2005) Weather, climate and tourism a geographical perspective. *Annals of Tourism Research,* 32(3): 571–591.

Mercer, J., Kelman, I., Lloyd, K., and Suchet-Pearson, S. (2008) Reflections on use of participatory research for disaster risk reduction. *Area,* 40(2): 172–183.

Montz, B.E., Galluppi, K.J., Losego, J.L., and Smith, C.F. (2014) Winter weather decision-making: North Carolina school closings, 2010-11. *Meteorological Applications*, 22(3): 323–333.

Mooney, C. (2016) What we can say about the Louisiana floods and climate change. *Washington Post*, August 15.

Moran, J.M. (2012) *Weather Studies Introduction to Atmospheric Science, 5th edition*. New York: American Meteorological Society.

Morss, R.E., Wilhelmi, O.V., Downton, M., and Gruntfest, E. (2005) Flood risk, uncertainty, and scientific information for decision-making: Lessons from an interdisciplinary project. *Bulletin of the American Meteorological Society*, 86: 1593–1601.

Morss, R.E., Demuth, J., and Lazo, J. (2008) Communicating uncertainty in weather forecasts: A survey of the U.S public. *Weather and Forecasting*, 23: 974–991.

Murray, I.R., Howie, C.R., and Biant, L.C. (2011) Severe weather warnings predict fracture epidemics. *Injury*, 42(7): 687–690.

Nichols, A.C. (2012) *How university administrators made decisions during National Weather Service tornado warnings in the spring of 2011*. Master's thesis, Department of Geography and Environmental Sustainability, University of Oklahoma.

Persson, M., Sundell, A., and Öhrvall, R. (2014) Does election day weather affect voter turnout? Evidence from Swedish elections. *Electoral Studies*, 33: 335–342.

Ranson, M. (2013) Crime, weather, and climate change (2014). *Journal of Environmental Economics and Management*, 67(3): 274–302.

Redniss, L. (2015) *Thunder and Lightning: Weather Past, Present, Future*. New York: Random House.

Ruin, I.J-D., Creutin, S., Anquetin, S., and Lutoff, C. (2008) Human exposure to flash floods – Relation between flood parameters and human vulnerability during a storm of September 2002 in Southern France. *Journal of Hydrology*, 361: 199–213.

Sabir, M., Ommeren, J., and Rietveld, P. (2014) Weather to travel to the beach. *Transportation Research Part A: Policy and Practice*, 58: 79–86.

Samenow, J. (2015a) Senate to hold hearing on improving weather forecast communication. *Washington Post*. http://www.washingtonpost.com/blogs/capital-weather-gang/wp/2015/04/16/senate-to-hold-hearing-on-improving-weather-communication/ (accessed July 23, 2017).

Samenow, J. (2015b) Wisdom on the state of weather forecasting and the embarrassing New York City blizzard forecast. January 28. http://www.washingtonpost.com/blogs/capital-weather gang/

wp/2015/01/28/wisdom-on-the-state-of-weather-forecasting-and-
the-embarrassing-new-york-city-blizzard-forecast/ (accessed
July 23, 2017).

Saulo, C. (2015) The hydrometeorological enterprise: The benefits of
partnerships. *Bulletin*, 64(1): March 2 http://public.wmo.int/en/
resources/bulletin/hydrometeorological-enterprise-benefits-of-
partnerships-0#sthash.jmi8Cl6I.dpuf (accessed July 23, 2017).

Schroeder, A.J., Gourley, J.J., Hardy, J., Henderson, J., Parhi, P., Rahmani,
V., Reed, K., Schumacher, R.S., Smith, B.K., and Taraldsen, M.J. (2016)
The development of a flash flood severity index. *Journal of Hydrology*,
April 8 (online).

Shepherd, M. (2016a) 5 reasons some were unaware of one of the biggest
weather disasters since Sandy. *Forbes Science*. August 16. http://www.
forbes.com/sites/marshallshepherd/2016/08/16/5-reasons-some-
were-unaware-of-one-of-the-biggest-weather-disasters-since-
sandy/#1db21c3a2520 (accessed July 23, 2017).

Silver, N. (2012) The weatherman is not a moron. *New York Times
Magazine*, September 7. http://www.nytimes.com/2012/09/09/
magazine/the-weatherman-is-not-a-moron.html?_r=0 (accessed July 23,
2017).

Skinner, M.W., Yantzi, N.M., and Rosenberg, M.W. (2009) Neither rain
nor hail nor sleet nor snow: Provider perspectives on the challenges of
weather for home and community care *Social Science and Medicine* 68
4 682–688

Strong C., Zhirui, Y., and Xianming, S. (2010) Safety effects of winter
weather: The state of knowledge and remaining challenges *Transport
Reviews*, 30: 677–699.

Sullivan, K. (2013) Restoring U.S Leadership in Weather Forecasting Part
2, before the Subcommittee on Environment House Committee on
Science, Space and Technology. U.S House of Representatives, June
26. http://www.legislative.noaa.gov/Testimony/Sullivan062613.pdf
(accessed July 23, 2017).

Thornes, J. (2008) Cultural climatology and the representation of sky,
atmosphere, weather and climate in selected art works of Constable,
Monet and Eliasson. *Geoforum*, 39(2): 570–580.

2

History of the Movement to Integrate Social Science Into Atmospheric Science

2.1 Early Weather Forecasting for Impacts

Historians use diaries to show how weather influenced everyday decision making. For example, historical diaries show how hot weather affected peoples' productivity and outlook. From the diaries of Mountstuart Elphinstone (the Scottish Governor associated with the government of British India from 1819 to 1927) and Lucretia West (the wife of the Chief Justice), Adamson learned how weather and climate affected their "everyday life" (Adamson, 2011). Sudan (2008) looks the interactions between the British East India Company and the British Royal Society related to "scientifically" manufacturing ice to help cope with the heat and monsoons. Sudan writes that the interest of Sir Robert Barker, who was stationed in India, had as much to do with the availability of comfortable refreshment as it did with the prospects of climate control, of making alien Indian weather English" (56). Fiebrich studied the history of surface weather observations that lead to the current automated networks, which include tens of thousands observations that come from amateurs, agencies, and private companies (2009).

Admiral Robert FitzRoy is credited with making the first daily weather predictions and calling them "forecasts" in 1854. He established the Meteorological Department of the Board of Trade to reduce ship sailing times using wind charts. Even though new evidence was being shown that weather charts could help to see instability between hot and cold air masses, most people still considered the weather "chaotic" and unpredictable (Moore, 2015). "When one Member of Parliament suggested in the Commons that recent advances in

Weather and Society: Toward Integrated Approaches, First Edition. Eve Gruntfest.
© 2018 John Wiley & Sons Ltd. Published 2018 by John Wiley & Sons Ltd.

scientific theory might soon allow them to know the weather in London 24 hours beforehand, the House roared with laughter" (http://www.bbc.com/news/magazine-32483678).

Fitzroy was motivated to create forecasts by the high number of lost lives at sea. Between 1855 and 1860, 7402 ships were wrecked off the coasts and 7201 people died. Fitzroy disseminated his forecasts using the telegraph ("a bewildering new technology" at the time). His scientific forecasts became very popular with people who were betting on horses, flower show organizers, and many others. As is still the case in 2017, FitzRoy faced strong criticisms when his forecasts were not perfect. His agency, now the UK Met Office, employs more than 1500 people and has an annual budget of more than 80 million pounds (Moore, 2015).

2.2   Historians and Weather

Since the focus of this textbook is the transformation of meteorological science to a discipline that is more inclusive of the social sciences, the discipline of history has some relevance. Dr. Roger Turner is one of several historians who studies weather as his area of expertise. How meteorologists have been perceived on television and how their image and content has changed have been studied by Turner and others, including Robert Henson (Henson, 2010b, 2010a). They look at how new forecasting techniques become the state of the art. Turner points out that weather is science's most popular genre. Lazo *et al.*'s (2009) surveys show 96% to 98% percent of American adults use weather forecasts regularly. More than 70% obtain forecasts from local television news at least once a day, one portion of the more than 300 billion weather forecasts Americans access every year. Even as local television stations are reaching smaller audiences and as people, especially younger people, rely more on their mobile devices than on network or cable TV, weathercasters remain celebrities, and stations offer high salaries to retain top-rated talents (Turner, 2009). "Viewers develop powerful bonds of trust with their local weathercaster, turning to the television for emergency news in the face of hurricanes, floods or tornadoes, and under happier skies, for advice on travel planning, car washing, and wardrobe construction" (Turner, 2009). Turner studies the processes that show how meteorological modeling was developed and grew to be standard practice. Turner has also studied changes in the ways that broadcast meteorology has become professionalized. He looked how weather broadcasting changed from having more of an entertainment focus to a more scientific bent while at the same time the gender of most broadcast meteorologists changed from women to men (Turner, 2009).

Vladimir Jankovic from University of Manchester, Cornelia Ludecke from University of Hamburg in Germany, and James R. Fleming from Colby College in the United States organized the 2008 Weather, Local Knowledge and Everyday Life meeting in Brazil. It was hosted by the International Commission for the History of Meteorology (http://www.h-net.org/announce/show.cgi?ID=155608) and the Proceedings were published (Jankovic and Barboza, 2009). The meeting, including many social scientists, explored how various disciplines conceptualize the evolution of "climatological citizenship" as it manifests itself in daily routines, rituals, perceptions, reactions to, and uses of the weather. Participants discussed the extent to which the weather matters in what an individual or a society does on a routine basis. The participants also gauged the depth of public assimilation of expert weather knowledge, media coverage, and decision making as it relates to atmospheric events, climate trends, and other forms of past and present "airmindedness." "They were particularly interested to unpack the ways in which the historical and contemporary actors and 'non-experts' experienced, remembered, predicted, ridiculed, feared and loved the weather."

General themes at this important initial meeting included the following:

- Non-meteorological approaches to weather and climate
- Non-institutional coping with extremes and disasters
- Science and local knowledge—methodological and theoretical agendas
- Everyday notions and practices about weather and health
- The cultures of blame, excuse, complaint and risk
- Seasonal knowledge
- Weather rites, festivities, taboos, and fetishism
- Public understanding of climate change and weather risk
- Weather and sports, outdoors, tourism, and travel
- Apparel, active wear, and extreme weather
- Weather buffs and storm chasers
- Weather and disability

- Indoor weather and indoor health
- Weather in daily speech and the media
- Visual, literary, olfactory dimension of atmospheric phenomena
- Aesthetics of atmospheric events
- Weather and religion and
- Weather and travel (http://h-net.msu.edu/cgi-bin/logbrowse.pl? trx=vx&list=h-nilas&month=0702&week=d&msg=pgRbFss2YOFR vrAy8DdD%2Bg&user=&pw=)

2.3   Weather and Society Efforts Build on Natural Hazards Research and Practice

Since the 1930s, geographers, led by Gilbert F. White, have had success in examining the physical and human influences on flood severity and mitigation (Burton, Kates and White, 1993). They consider the whole process from warnings through recovery from extreme weather. Still, there has been little crossover between hazard researchers and studies of weather. For decades, many scholars have recognized that social sciences would inform the work of meteorologists and hydrologists. Natural hazards research and practice understood the limitations of single-discipline approaches to challenges of reducing losses from floods, tornadoes, and other hazards. Sessions at professional meetings, research proposals, and scholarly publications routinely mentioned the importance of applying what is learned from research to real-world problems to reduce losses and improve policies in sustainable ways (e.g., Montz and Gruntfest, 2002). White and his colleagues showed that even though so much more is known about vulnerability and hazard characteristics, losses continue to increase because of unwise land decisions, inability to learn from experience, lack of disincentives for developing in vulnerable areas, and other factors (White *et al.*, 2001).

Early work in natural hazards highlighted the need for integrating several disciplines and pointed out the necessity to learn from disasters. White and Haas's (1975) *Assessment of Natural Hazards* called for comprehensive monitoring and understanding of the risks and physical dimensions of the hazards. By the 1999 Second Hazard Assessment (Mileti, 1999), human intentions and decisions, rather than worsening storms, were blamed for increasing vulnerability. Mileti (1999) called for building "disaster-resistant communities" with attention to wise long-term land-use planning to reverse increasing disaster damages. Despite numerous calls for interdisciplinary collaboration, notably from National Science Foundation research funding initiatives and National Research Council reports, most academic departments remain in silos, separated by traditional disciplinary boundaries (Committee on Disaster Research in the Social Sciences, 2006).

The National Science Foundation has funded some interdisciplinary research projects that include social scientists and atmospheric scientists through its Hazards Science, Engineering and Education for Sustainability (SEES) program (http://www.nsf.gov/awardsearch/advancedSearchResult?ProgEleCode=8087&BooleanElement=ANY&BooleanRef=ANY&ActiveAwards=true&#results) and its Dynamics of Coupled Natural and Human Systems (CNH) program (http://www.nsf.gov/news/news_summ.jsp?cntn_id=129178).

CNH's stated goals embrace a multidisciplinary approach. Dr. Susan Cutter, the Director of the Hazards and Vulnerability Research Institute at the Department of Geography at the University of South Carolina, made the following comment related to all disasters, including weather-related disasters:

> We must become more proactive in our approach to disaster losses by increasing our resilience. For example, individuals and communities must realize they are their own first lines of defense against disasters, and the decisions they make about where to build, how much to invest in hazard mitigation and what land use, zoning laws and building codes to use will influence their ability to respond to and recover from disasters. Present disaster policies lack mechanisms for fostering disaster resilience, instead privileging short-term gains and interests over long-term commitment and investments for the future. As a nation and a state, we lack a sustained commitment to reducing disaster risks... If we could divert some of the billions of dollars now used for disaster response or recovery and instead invest these resources today to build resilient communities for tomorrow, we would save not only lives but money as well. Building resilience to disasters is everyone's responsibility, not just a responsibility of governments, political parties or the private sector. We need to move beyond responding to the crisis of the moment and work collaboratively to manage our risks and

enhance resilience for all sectors and all communities. Disaster resilience is not something we think about in our two-, four- or six-year election cycle window. It needs to become the basic fabric of our communities and our everyday lives, as we never know when or where the next disaster will occur (Cutter, 2013).

Dr. Dennis Mileti's book *Disasters by Design* argues that human activity increases hazard vulnerability (Mileti, 1999). Other classic hazard textbooks have the same message (Tobin and Montz, 2015). Even though the boundaries are blurry (and vulnerability and resilience are important topics), this book steers away from natural hazard examples that may be covered in "disaster" textbooks and delves more deeply into severe weather examples.

The hazards literature provides many examples of social and physical scientists working together to mitigate the impacts of weather hazards. Excellent references can be found through the clearinghouse for hazards research: the Natural Hazards Research and Applications Information Center at the University of Colorado in Boulder, Colorado. The Center's website (www.colorado.edu/hazards) provides online newsletters for keeping up on conferences, other hazard research institutes, research funding, and job possibilities. Numerous universities offer courses on natural hazards or disasters, and many universities offer courses that consider issues at the intersection of weather and society. Some of the universities and the departments that offer these courses include East Carolina University (Geography), University of South Carolina (Geography/HVRI), the University of Colorado at Boulder (Sociology/Geography), Louisiana State University (Geography and Anthropology), the Disaster Research Center at the University of Delaware, the University of Arizona (Geography), Arizona State University (School of Sustainability), the Hazard Reduction and Recovery Center at Texas A&M University (Psychology, Sociology, Geography, Political Science), Oklahoma State University (Political Science/Emergency Management), University of New Orleans (Sociology), George Mason/Yale/Cornell/Kentucky and Colorado State, University of Southern California (CREATE), University of Georgia (Geography), the University of Oklahoma (Geography), Texas State University (Geography), Northern Illinois University (Geography), Columbia University, University of North Carolina, University of North Texas, University of Missouri, Kent State University, University of West Virginia, University of Akron, and Eastern Illinois University (Gottlieb and Klockow, 2013).

2.3.1 Efforts to Add Social Dimensions to Solving Weather Problems

The National Center for Atmospheric Research (NCAR) in Boulder, Colorado hosted the Environmental and Societal Impacts Group (ESIG) for about 30 years starting in 1970. Dr. Mickey Glantz was the first director, and he led the group for more than two decades. In 2004, NCAR reorganized and ESIG changed into the Institute for the Study of Society and Environment (ISSE).

A 1997 ESIG workshop on the social and economic impacts of weather calling for research on weather decision making (http://sciencepolicy.colorado.edu/socasp/weather1/index.html) was held in Boulder, Colorado. Numerous social scientists, including sociologist Dr. Dennis Mileti, psychologists Dr. Tom Stewart and Dr. Gary McClelland, and economist Dr. Alan Murphy participated in the workshop along with many industry experts from the oil and gas, electric power, surface transportation, agriculture, aviation, and insurance industries. The participants shared examples of the effects of weather that decision makers face in their day-to-day operations and the potential value of improved weather information and information use. Calling for new social science and weather research, participants cited examples of how research could reduce the economic impacts of severe weather. Workshop participants identified a set of social science methodologies to research the use and value of information to decision makers (Pielke Jr., 1997). Most social science considerations at the workshop were strictly economic, however, rather than social science in a broader sense.

Workshop participants discussed many tools that can be used to show the societal value of improved forecasts and to contribute to improved use of weather information by decision makers. Pielke Jr.'s 1997 article illustrated how the model of Research-to-Operations was scientifically driven rather than societally driven. Pielke wrote:

> Because society undergoes constant change, our problems evolve. Consequently, there is a constant need to remain vigilant as to the nature of the relationship between scientific research and society. In many cases, securing the use of science to aid in addressing societal problems is a difficult and challenging task. However, it is a surmountable challenge that will in the long run prove to be beneficial to both the institution of science and the broader society of which it is a part... The

atmospheric sciences community must be responsible to ensure that the problems it seeks to address are well defined, that social scientists and users are included in the problem definition process, and that the lessons of experience are distilled and disseminated (Pielke Jr., 1997:263).

Pielke went on to say, "A perfect forecast is of no value if it is unavailable to or unusable by a decision-maker... Consequently, attention must be focused on the use and value of weather information in parallel to ongoing efforts to improve the quality of the information" (Pielke Jr. and Carbone, 2002). Many American Meteorological Society and National Weather Association committees, reports, and publications have addressed this issue since 1997, including the Societal Impacts Board of the American Meteorological Society.

The 1997 workshop led to a 2003 proposal to the U.S. Weather Research Program and the funding of social science research programs at the National Center for Atmospheric Research (NCAR) in Boulder, Colorado. Most of the research questions and priorities that were identified in the 1997 workshop report are still relevant in 2017.

2.3.2   Weather and Society * Integrated Studies—WAS*IS

In 2005, the U.S. Weather Research Program funded the Societal Impacts Program at NCAR. A "methods" workshop was part of the initial NCAR Societal Impacts program proposal. The Weather and Society * Integrated Studies (WAS*IS) program was established to satisfy the need for the methods workshop (Figure 2.1). The acronym shows the change from what WAS to what IS the future of integrated weather research and practice. When WAS * IS began as a movement in 2005, it had six main original goals:

1) To lay the groundwork for conducting interdisciplinary work by learning new strategies and addressing typical challenges including: developing strategies for success, recognizing challenges, and acknowledging the long time frame and sustained commitment necessary to do integrated work.
2) To teach basic tools and concepts as essential components of integrated weather-society research and applications. Since most WAS*ISers are meteorologists, most of the workshop curriculum

Figure 2.1 Button for WAS * IS (Weather and Society * Integrated Studies). *Source:* WAS*IS.

focused on introducing social science concepts in different weather contexts.

3) To learn about effective integrated research and applications using real-world examples.

4) To build a large community of dedicated meteorologists and others who rely on formal and informal networks including Facebook, Twitter, chats, and blogs as main sources of data and information.

5) To identify and pursue research, application, and educational opportunities for integrated weather and social science work.

6) To improve and facilitate the ongoing relationships among practitioners, researchers and stakeholders in meteorology and the social sciences (Demuth *et al.*, 2007:1730-1731).

As of 2017, there are 277 official WAS*ISers. Most are in the U.S, but WAS*ISers also hail from the Netherlands, Germany, France, Australia, Canada, New Zealand, Puerto Rico, Guatemala, Trinidad, Cuba, Antigua, Mexico, Costa Rica, St. Vincent, Barbados, Jamaica, and Granada. Most WAS*IS workshops took place in an intensive eight-day, in-person format. This structure helped develop deep and lasting professional and personal relationships among the participants.

The WAS*IS experiences show that when meteorologists and social scientists work together, they provide results and products that are greater than the sum of their individual efforts. In academia, government, the private sectors, and non-governmental sectors recognition is growing that integrated thinking and problem solving across disciplinary boundaries are required in meteorology and hydrology to effectively address the issues of societal impacts. Within the weather world, many early career physical and social scientists and engineers seek to learn about concepts, tools, questions, and policies related to more than one discipline, but most academic departments, especially in meteorology and hydrology, are too narrow to allow much leeway for electives outside of their defined discipline.

WAS*IS addresses the needs of a growing community of people who recognize that tackling today's complex problems, including reducing losses from weather events, calls for interdisciplinary cooperation and multi-disciplinary approaches. The WAS*IS movement is part of an initial leap toward changing the culture of atmospheric science to be more inclusionary. It includes new perspectives from a wide variety of specialties, taking into account concepts of resiliency, societal impacts of weather, and new ways of thinking about forecast verification. The new outlook requires a leap into other points of view, being open to new ways of seeing weather and its impacts. WAS*IS presents ideas that are being developed by a community of scholars and practitioners dedicated to changing the old outlook.

Unfortunately, WAS*IS funding ended in 2012 (the website is available as archives: www.sip.ucar.edu/wasis). As of 2017, however, there are 1000 people on the WAS*IS Facebook page where conversations consider new career opportunities, share new research findings, debate controversies related to recent warnings, discuss social media dissemination of information, talk about public and private weather partnerships, and bring up many other topics. In 2015, Castle Williams and Minh Phan, two graduate students, initiated a Facebook page for WAS*IS students. Even though they began their careers after the end of the formal workshops, they are carrying the torch and bringing in new students eager to be part of the movement from WAS to IS. Their page is a forum for thoughtful dialogue at the crossroads of weather and society matters, including conflicting information related to forecasts from government and media sources and the implications of a TV meteorologist offering to embarrass employers who insisted that their employees come to work even during record Houston flooding in April 2016 (https://www.washingtonpost.com/news/capital-weather-gang/wp/2016/04/19/we-will-expose-that-person-meteorologist-threatens-bosses-who-force-people-to-commute-in-flood/).

2.3.3 Integrated Warning Team Meetings

One successful WAS*IS spinoff has been the Integrated Warning Team (IWT) meeting. Dozens of these meetings have been held all over the United States. The meetings, usually hosted by the NWS, have a regional focus and bring together all partners including media, emergency managers, transportation departments, local businesses, private weather forecasters, and others to discuss ways to improve collaborations before, during, and following severe weather. The meetings range from a half-day to three days and focus on local and regional risks and challenges. Some regions have hosted annual or biannual IWT meetings and show progress in communication strategies and improved trust and interaction. In Colorado in 2012, a state-wide IWT was held, and more than 75 people attended. Appreciation of the benefits of positive collaboration and renewed trust were noticeable during the severe fires that occurred later that year. Kansas City, Missouri and Atlanta, Georgia are examples of communities where IWT's have been held with measurable results.

2.3.4 American Meteorological Society Summer Policy Colloquium

For 10 days every summer, the AMS Summer Policy Colloquium brings a select group to Washington, DC for an intense, 10-day immersion in science policy. Atmospheric, social, and policy science graduate students, faculty, and professionals hear from dozens of prominent experts and build long-lasting professional networking connections and friendships. The Summer Policy Colloquium is a career-shaping experience. The Colloquium "arms tomorrow's leaders with expertise in the policy process so that the science community will be more engaged with decision makers, helping to ensure that society's policy choices take full advantage of available scientific knowledge related to weather and climate" (https://www.ametsoc.org/ams/index.cfm/policy/summer-policy-colloquium/).

2.3.5 Social Science Woven into Meteorology (SSWIM)

In 2006, the U.S. National Weather Center building opened in Norman, Oklahoma. About 600 meteorological researchers, students and practitioners work in the building. Many federal research and operations agencies are located in the National Weather Center, including the National Severe Storms Laboratory, the Hazardous Weather Testbed, the Warning Decision Training Branch, the NWS Storm Prediction Center, and the Norman Weather Forecast Office, and the University of Oklahoma's School of Meteorology. With organizational help and leadership from National Weather Center scientists, a partnership between the NOAA and the University of Oklahoma Social Science Woven into Meteorology program was created at the National Weather Center to bring in some social scientists to work with meteorologists. The new acronym SSWIM (for Social Science Woven into Meteorology) emphasized that the group was designed to weave social science into the fabric of the National Weather Center and not to simply tack it on.

During the 2008-2012 SSWIM program, a part-time director, a social science post-doctoral scientist, and graduate students committed themselves to integrating social science and meteorology. Demand for the post-doc and graduate student positions was strong and highlighted excellent early-career scientists who were willing to be pioneers in this new effort. Dr. Heather Lazrus, an environmental anthropologist, was the post-doctoral scientist. Several early-career pioneers shared the graduate student positions.

Many physical scientists in the National Weather Center reached out to the SSWIM team, expressing curiosity about how to start collaborations related to new warning technologies, improved understanding of how people use NWS watches, behavioral verification for flash flood forecasts, and other concerns. The SSWIM team was enthusiastically received as new colleagues in the National Weather Center and numerous meetings were held. Most of these meetings could be characterized as cross-cultural events where each person needed to provide details on their work and what they were trying to accomplish as well as sharing their perceptions of how social science could help them better reach their objectives.

SSWIM was successful at beginning new integrated efforts with the U.S. Storm Prediction Center, the Warning Decision Training Branch, the Multi-function Phased Radar project, the Warn on Forecast group, and the flash flood research group of the National Severe Storms Laboratory. SSWIM developed research and built capacity by increasing appreciation of the value of qualitative as well as quantitative research approaches, including archival, ethnographic, and participatory methods. SSWIM created partnerships with public, private, and academic sectors, including students, practitioners, and policy makers across the spectrum of stakeholders.

There were some stumbling blocks and frustrations. In some cases it seemed that meteorologists thought that by merely talking with a social scientist, they would learn how the public used their forecasts and what was necessary to improve them. Others thought there would be a magic social science formula that could be applied to answer their numerous questions about what made forecasts effective and what would motivate people, the public, to take appropriate actions in the time that they had. Administrators frequently asked for data that could be directly ingested into current physical science models as the way to integrate social science. Other meteorologists thought that all social scientists were economists and that SSWIM could quickly place a value on the services they were providing.

The SSWIM team patiently explained that social science research was necessary to give the answers they needed, but SSWIM faced political headwinds and did not have the funds to comprehensively develop the research projects to meet the challenges. Changing the process and getting the answers they needed required more than a few meetings. Many meteorologists wanted to know the extent of the societal impacts and had unrealistic expectations at first. The process of integrating social science into meteorology is time consuming and not inexpensive.

SSWIM's four years, from 2008-2012, were filled with program initiation work, building partnerships, academic, governmental and public outreach, and proposal writing. The SSWIM group generated interest and more possibilities for collaborative work between social scientists and meteorologists and conducted research that provided the foundation for literature in social science that is central to progress in integrated weather research and practice. Stephanie Hoekstra and Amy Nichols wrote Geography Master's theses at the University of Oklahoma that show how tornado warnings are interpreted by university emergency managers and by public school administrators (Hoekstra, 2012; Nichols, 2012). They had strong backgrounds in meteorology but recognized that social science study would help them address the challenges they found most compelling at the intersection of weather and society. As of 2017, some researchers at the National Weather Center are working on social science aspects of weather, including political scientists Dr. Joe Ripberger, Dr. Hank Jenkins Smith, and Dr. Carol Silva at the Center for Risk and Crisis Management (http://crcm.ou.edu/). Hoekstra finished her Ph.D. in coastal resource management at East Carolina University, writing her thesis on the decisions made by officials during the 2012 Hurricane Sandy (Hoekstra, 2015).

Building partnerships takes time and can feel awkward and uncomfortable. Building these networks is difficult, messy, and very time consuming. No one in the weather enterprise is specifically tasked with the job of building these relationships. And the development of networks is not a product that is easy to measure in performance metrics as part of meteorologists' jobs.

Yet learning to talk with each other and trusting each other are prerequisites to addressing weather and society issues. The path is winding. People with one skill set do not easily recognize and appreciate the ways other partners do their work. However, with time and discussions about the impacts of weather events and considerations of new policies and practices, many meteorologists regularly interacted with students and researchers within SSWIM's four-year period and these interactions had lasting effects.

Trust between collaborators is difficult to measure, but it is significant that social scientists are now comfortable in working relationships as parts of teams with engineers and physical scientists. When SSWIM began in 2008, it was unusual and somewhat unsettling for social scientists and physical scientists to meet in the Hazardous Weather Testbed or in the Development Lab. With time, each person claimed a rightful seat at the table. SSWIM and its partners needed to overcome problems of finding a common language and even compatible, rather than parallel, conceptual models. The following sections highlight some examples of research and hazard events with both social science and meteorological/climatological considerations.

2.3.6 VORTEX-Southeast

The Verification of the Origins of Rotation in Tornadoes EXperiment-Southeast (VORTEX-SE) is a research program to understand how environmental factors characteristic of the southeastern United States affect the formation, intensity, structure, and path of tornadoes in this region. VORTEX-SE will also determine the best methods for communicating forecast uncertainty related to these events to the public and evaluate public response. This program that began in 2016 includes a major social science component. Vortex-SE has three major research emphases: Observing and modeling tornadic storms and their environments ($2,885,000), improving forecast models ($857,000) and addressing risk awareness and response and tornado damage mitigation ($741,000). The social science research efforts are the following:

1) addressing interconnections between the built and natural environments through post-event damage surveys; 2) tornado warning response in the Southeast: Advancing knowledge for action in Tennessee; 3) complacency and false alarms in tornado affected communities; and 4) Understanding the current flow of weather information and associated uncertainty, and their effect on emergency managers and the general public. (http://www.nssl.noaa.gov/projects/vortexse). It will be exciting to follow the progress of this major experiment as the social scientists and atmospheric scientists work in the field together to better understand tornadoes in the southeastern United States.

Research Spotlight Box: Doing Social Science Reveals Challenges to Accepted Assumptions

Research findings often challenge the conventional wisdom or commonly held assumptions. For example, it is a commonly accepted fact that it is an indicator of progress when a county buys a siren system that offers the opportunity to narrow the geographic zone for sounding the sirens, thereby notifying only the people where a warning is in effect rather than sounding all sirens throughout the entire county at one time.

Weather researcher Cedar League's work, conducted at University of Colorado at Colorado Springs, had surprising results. Just because a community has purchased a siren system that allows emergency managers or law enforcement officials to notify sections of the county threatened by severe weather, in many cases, they continue to sound the sirens for the entire jurisdiction even when many parts of the jurisdiction were not at risk.

NWS forecasters issue tornado warnings when there is a radar signature or a visual confirmation of a tornado on the ground. Many emergency managers in towns with sirens do blow the sirens as soon as the NWS warning is issued. Before 2000, most siren systems were established at the county level. This structure was in keeping with the county-wide storm warnings issued by the NWS. However, now the NWS issues polygon warnings that include parts of counties. Some new siren systems have "zones," so that if a storm occurs in the northeast corner of a county and the storm is heading east, then sirens in the western part of the county would not need to be activated.

League conducted a survey and focus groups and asked emergency managers from Texas and Oklahoma about their siren operating procedures for tornadoes. To reach a large contingent of emergency managers, League attended emergency managers' professional association meetings where she could connect with hundreds of emergency managers. Many emergency managers were willing to help her by participating in focus groups and answering a series of questions. League's research aimed to help the new CASA radar implementation team understand what types of weather information are most valuable to emergency managers and to find out what the emergency managers are currently doing with the warning information they do have.

League's findings show that some EMs with the ability to warn by sub-region are not employing this capability because they are concerned about missed events and they prefer to practice a "better safe than sorry" policy. Some are blowing the sirens and warning entire counties even when only a small portion of the county is included in the NWS warning. In 2016, Oklahoma City implemented zone tornado warnings (Allen, 2016). Implications of this work are that technological, policy, and training interventions will be required to take advantage of warning by sub-region. When sirens are activated for many storms that result in no tornadoes in most locations, the practice can reinforce public and EM concerns about "over-warning," which leads the public to ignore all warnings.

Through open-ended questions in the focus groups, League also learned that NWS warnings are not the only prompts that emergency managers use to sound county warning sirens. When asked if their jurisdiction always warns when a NWS warning has been issued, 64% of the Oklahoma emergency managers and 61% of the Texas emergency managers said "yes." Furthermore, 79% of Oklahoma and 60% of Texas emergency managers said they would warn the public about a tornado threat when a NWS warning had not yet been issued. These percentages indicate that many emergency managers also rely on their spotters and other information besides the formal NWS warnings (League *et al.*, 2010; League *et al.*, 2012).

These two major findings from the focus groups and surveys of emergency managers must be taken into account by software developers, radar engineers, and forecasters if they want their new "improved" systems with greater temporal and spatial specificity to truly help emergency managers effectively warn the public about approaching severe weather.

2.4 Physical Science Myths Related to Climate and Applicable to Weather

Dr. Diana Liverman, shown as Figure 2.2, is a professor of geography at the University of Arizona. She has decades of collaborative research working at the intersection between natural and social sciences. Her research focus has been on global change with an emphasis on climate change. There are clear parallels between the difficulties of integrating social science into climate studies and into weather studies. In 2011, she summarized myths that she says "make it difficult to collaborate with colleagues in the natural sciences." Six of the myths Liverman identified are particularly relevant to the social science/weather integration issue and are paraphrased below (Liverman, 2011).

Myth 1: Physical science alone has the answers to the question of how to reduce loss of life and property from severe weather. Meteorologists worry about droughts due to the lack of rainfall. However, fieldwork shows understanding impacts requires research on economic conditions, farming practices, and other socio-economic factors at least as much as on meteorological factors.

Myth 2: The "loading dock" model, or the knowledge or information deficit model assumes that the reason the public and decision makers are confused about weather and warnings is because *they just*

Figure 2.2 Professor Diana Liverman. *Source:* S. Meckler. Reproduced with permission from Institute of the Environment.

don't have enough information. Many scientists assume that the way to make research useful to stakeholders is to give a power point talk at a workshop or put your results on a nice web site. Social science research shows us that this is not true. To effectively connect with your audience you need to understand its needs, the ways people frame the weather issues and the importance of discussing their needs with them to coproduce understanding and improve decisions (Liverman, 2011).

Myth 3 has three parts that sometimes conflict. People are irrational about risk, attitudes explain behavior, and humans are predictable. The NWS and other meteorological agencies think that if they simply improve their forecasts to be more accurate and more precise that people will do the right thing in the time they have to reduce their vulnerability. Improving forecasts is a good thing, but it is not the only thing that guarantees that vulnerable people will be motivated to change their behavior. Making a decision about taking an action in severe weather is more complex than just responding to the stimulus of an official warning. Although it may seem irrational to go out when severe weather warning are in effect, parents sometimes are compelled to pick up their children, and workers are insistent upon going to work because they fear repercussions if they don't get to work on time.

Ruin *et al.*'s (2008) and League's (2009) research shows that many people do know there are flood warnings in effect and that the low water crossings they pass through might be flooded, but they decide to go forward because they need to go to work. They believe that the risk of losing their jobs is greater than the risk of losing their life in a flood. For these cases, the key to reducing vulnerability and changing behavior is not better or more information or even better dissemination, it is changing the employer's insistence that people get to work on time regardless of the weather conditions. The Lower Colorado River Authority in Austin, Texas is an employer that has changed its policy explicitly because of these social science research findings. Employees can call in when the roads might be flooded, and they will be allowed some flexibility to change their route or to come in later after the flooding subsides.

There are many documented reasons why people behave as they do. While the behaviors may appear to be irrational to many physical scientists, closer looks reveal common sense reasons for why people do what they do when faced with severe weather. Evacuation prior to hurricane Katrina's arrival in New Orleans seems like it was the only "rational" choice based on the forecasts. More than 80% of the population did evacuate. However, some people were unable to evacuate for many reasons including having no transportation, and not knowing a place that would accommodate them and their pets.

Myth 4: There is a "general public." Meteorologists must recognize that there is no one-size-fits-all solution to getting people to change their behavior appropriately in the face of severe weather or warnings. Creative, new, integrated research and numerous experiments are necessary to find out what techniques work and in what contexts. There will not be a single solution that leads to more people doing the right thing in the time they have, whether it's going to the storm cellar, evacuating the coast, or changing their route to avoid low water crossings.

Myth 5 collects two conventional views and dismisses them: a) There is a presumption that social science is too political, cheap, applied, lax, and qualitative to be the equal of the natural sciences in weather research. Many physical scientists undervalue the methods and the depth of social science. b) Physical scientists act as if social science is not science. Some atmospheric science academic departments fear that when physical scientists engage with social scientists it dilutes their "science." They act as if there is one option for science that only considers physical characteristics of the atmosphere. There is a reluctance to take social sciences into account.

Myth 6 is that social science research requires minimal funding. What's involved in social science research? While expensive funding for satellites to collect data are not required, that does not mean social science is inexpensive. Social scientists need funds for personnel and related costs for conducting mail surveys or focus groups. They need field research funds. As with atmospheric science research, more than one case study is necessary to develop datasets that have generalizable meaning and to get the observations that are necessary to be able to have confidence in the results.

Dr. Liverman's insights reinforce the need for the weather and climate communities to work together more closely. The separation is more pronounced in the United States than it is elsewhere. In Canada, Australia, and New Zealand, the continuum between weather and climate is appreciated and work is more seamless.

2.5   Meteorological Aspects of Non-Meteorological Hazards

Many hazards are not meteorological per se but have meteorological aspects. Volcanic eruptions and heat waves are two examples. Volcanoes are geophysical events. Yet during April and May 2010, the eruptions of Eyjafjallajökull volcano in Iceland led to fear that the volcanic ash would damage aircraft engines. The eruptions caused the largest air-traffic shutdown since World War II. On April 16-17, 2010, 32,000 of Europe's usual 56,000 daily scheduled passenger flights were cancelled. By April 21, 95,000 flights had been cancelled. The extent of disruption from the eruption depended upon the weather. The ash from the volcano in Iceland was carried by the wind (Donovan and Oppenheimer, 2011:1). These events pose the question as to why technocratic nations in northern Europe seemed so surprised and poorly prepared for an event that was anticipated in the scientific community.

The European volcanic ash example represents the tension at the weather–society interface. For years prior to the 2010 ash disruption, there were daily ash cloud trajectory model runs using hypothetical wind patterns from Iceland's volcanoes. "These preparations demonstrate that the aviation hazard posed by Icelandic volcanism has been recognised by scientists, operational meteorological institutes and the aviation authorities for many years" (Donovan and Oppenheimer, 2011:5). Since the 2010 eruptions and disruptions, new guidelines for international collaborations between agencies, scientific organizations, and airlines have been worked out.

Meteorology is not the first discipline to consider integrated approaches. Geophysicists have included historians and anthropologists at their biannual international "Cities on Volcanoes" meeting for more than a decade. Volcanologists recognized that learning what historians and anthropologists know about how people prepare for and rebuild after eruptions provides key data to help understand the geophysical composition of the volcano, the types and dates of the historic eruptions, and how communities have dealt with the hazard. "Cities on Volcanoes" established "a link between scientists, society and decision makers to develop integrated risk management, by encouraging the exchange of experiences and knowledge regarding volcanic phenomena, its impact on society and environment, and by improving risk reduction measures" (http://www.citiesonvolcanoes9.com/en/).

Environmental health is another area where meteorologists, epidemiologists, climatologists, and others work in inter-disciplinary ways with social scientists, including anthropologists, geographers and political scientists. Reducing loss of life from heat waves has been a recognized success of integrated approaches. A heat wave is not just a matter of a particular high temperature or a number of days where high temperatures are experienced.

Based on a study of summer mortality, a Philadelphia, Pennsylvania heat wave is based on "the number of consecutive days the air mass has been present, maximum temperature, and the whether the oppressive air mass occurs early or late within the summer season (Kalkstein *et al.*, 1996:1521). Forecasters can provide two days lead time before the oppressive heat arrives. This gives the Philadelphia Department of Public Health time to issue health alerts and warnings to the media. Promotion of the "buddy" system encourages neighbors to check on each other. Other actions to reduce vulnerability include the suspension of utility cut offs, more visits by public health workers, increased staffing at emergency rooms and air conditioned facility operation (Kalkstein *et al.*, 1996).

Oppressive heat is deadly. Extreme heat killed more than 700 people in Chicago, Illinois in July 1995 (Klinenberg, 2003). Summer heat waves killed more than 50,000 people in Europe in 2003, mostly in France and Italy (Larsen, 2006; Semenza *et al.*, 1996; and NOAA 2012). The death rate was an average of 60 percent higher than usual because of the 40 °C or 104 °F temperatures (Boseley, 2015). More people die in the United States annually from extreme heat than from hurricanes, lightning, tornadoes, floods and earthquakes combined (NOAA, 2012). A heat vulnerability index (HVI) was developed from a suite of eight demographic, health and land use variables. In Georgia, in the United States, a study of heat wave vulnerability focused on rural high-vulnerability counties, rather than the urban areas (Maier *et al.*, 2014). They point out that social isolation, population of elderly and poor health were the dominant factors in the more rural counties. While the temperatures in that heat wave were not as extreme as the Russian heat wave, the night-time low temperatures in the 2003 heat wave were considerably higher. This condition tends to add to heat stress and causes a higher death toll. Heat waves are considered "silent killers: affecting mostly the elderly, the very young or the chronically ill" (Larsen, 2006).

In summer 2010, a severe heat wave affected Russia killing 7000 people in Moscow and up to 15,000 people country-wide. Record high temperatures were reached on more than six days. The problems the heat caused had major socio-economic and public health impacts. The heat and dryness led to severe peat and forest fires. These fires had major impacts on air quality in the metropolitan area, so city dwellers could not open their windows to get fresh air because the smoke made the air quality so bad. Carbon monoxide levels rose to as high as 6.5 times the admissible maximum level (McElroy, 2010). Planes could not land or take off from the airports because the thick smoke reduced visibility below safe levels, and this hindered the ability for planes to be used to fight the fires, too (Masters, 2010). Dealing with the impacts of volcano eruptions and heat waves show the necessity of employing inter-disciplinary collaborations between meteorologists and social scientists.

2.6   Research That Evaluates What People Did When They Heard a Warning: Hurricane Ike 2008 Case Study

What did people do in Galveston, Texas when they heard the National Weather Service Call to Action Statements in 2008 for Hurricane Ike? Drs. Morss and Hayden went to Galveston, Texas to conduct interviews a couple weeks after the impact of Hurricane Ike. They asked people to comment on the Call to Action statements issued by the National Weather Service as part of the Hurricane warnings. The Call to Action statement was part of the hurricane warning.

Before the storm's impact, the unusually severe and specific warning was: "Neighborhoods that are affected by the storm surge ... and possibly entire coastal communities ... will be inundated during the period of peak storm tide. Persons not heeding evacuation orders in single-family, one- or two-story homes may face certain death. ... Widespread and devastating personal property damage is likely elsewhere." The National Weather Service advisory also warned that in some places, floodwaters could be as much as 9 feet deep more than 1 mile inland.

Drs. Morss and Hayden found that about 70% of interviewees (35) said they heard this statement before Hurricane Ike hit the coast. More than two-thirds of those who remembered where they heard the statement said they heard it from television, corroborating the earlier discussion about the importance of television as a source of information about an approaching hurricane (Morss and Hayden, 2010). Other sources of information that people identified included radio, the Internet, and other people. Of those who heard the statement, 14 people expressed opinions that were generally positive, saying they believed the statement or it increased their awareness of the risk.

Thirteen people expressed negative opinions, saying they did not believe the statement or thought it was exaggerated, too extreme. Four people expressed mixed opinions, and the remainder could not be categorized. Most of those who had heard the statement recognized it immediately, and many had a strong opinion about it. Positive phrases used to describe the statement include "blunt...effective," "correct," "to the point," "scared you to death," and "people who didn't heed were foolish." Negative phrases include "harsh and over reactive," "overblown," "ridiculous," "humorous," "stupid," "rude," and "not appropriate." Some noted the trade-off between scaring people too much and convincing people to leave. One interviewee thought the statement was too extreme, but he understands some people do not listen unless the warnings are strong. Another interviewee said that although the statement was very true and people should leave, it added too much stress and was very disturbing. Overall, some said they took it seriously, and others said they did not.

The "certain death" statement made a strong impression on a number of people. How did the statement affect their decisions leading up to Ike? More than two-thirds (25) of the interviewees who heard the statement said that it did not affect their decision to prepare or evacuate. This can be partially explained because many people had already left their residences by the time the statement was issued. Of the 10 whose decision was affected, 8 said it helped them decide to evacuate, and 2 said it reinforced the decision they had already made to evacuate. One interviewee said the statement helped her talk her husband into leaving. Another interviewee said that it made her realize that the storm would be bigger than expected. One person said that the statement was the "biggest factor in being convinced to evacuate" while another person simply said, "That's why I left."

These results suggest that the NWS statement motivated at least a few people to leave and thus may have saved lives. However, some people who had strong negative opinions about the statement indicated that it might decrease their sensitivity to communication of future hurricane threats. The researchers found that responses from non-evacuees include, "Such statements make him lose confidence; another said the statement made him want to stay 'to show them he's not going to die.'" These responses illustrate the importance of learning how to effectively communicate the risk associated with a threatening hurricane in ways that motivate people at high risk to take action without substantially reducing future response.

Many interviewees heard the "certain death" statement issued by the National Weather Service, and it helped convince several to evacuate. However, others had strong negative reactions to the statement that may negatively influence their response to future warnings. Their findings show that empirical studies of how intended audiences obtain, interpret, and use hurricane forecasts and warnings can be very valuable for designing effective hurricane risk messages and dissemination strategies. Their study was not comprehensive but does highlight the value of conducting future studies before generalizing about the advantages or disadvantages of Call to Action statements. "Concerned about the projected storm surge, National Weather Service (NWS) forecasters issued statements warning that some coastal residents will or may face certain death." While evacuation decisions are sometimes viewed or modeled as one-time, yes–no decisions, in reality, they are often complex and evolving. As discussed in Gladwin and Peacock (1997) and Gladwin *et al.* (2001), Morss and Hayden found that risk perception, experience, evacuation orders, forecasts, environmental cues, household interactions, and resources can all interact over a period of time as people decide whether to evacuate, when, and how (2010).

One lifelong Galveston resident who didn't evacuate had stayed in place during previous hurricanes. Because the interviewee's area had not flooded in his lifetime and his home's main floor is elevated, he did not anticipate flooding despite the forecasts. However, the warnings concerned his teenage daughter, and she and his girlfriend convinced him to evacuate. They packed to leave, but he had only $50 and was unable to obtain additional cash because banks were already closed, and no one else would cash his two-party paycheck. Also, he anticipated difficulty finding a hotel to stay in with his five pit bull dogs. His family said that if he stayed, they would stay, so he purchased supplies and they all stayed. Although their main floor did not flood, the man had to rescue his dogs from the first-floor garage, and he and his family suffered emotional trauma from experiencing the storm. This suggests that despite recent efforts to assist hurricane evacuees with pets (Burns, 2008), pets remain a factor in some people's evacuation decisions (Morss and Hayden, 2010:180). These are the types of human factors that meteorologists are unlikely to consider when they craft weather warnings. The insights offered by social science research can vastly improve how we prepare for severe weather and the effectiveness of such warnings when severe weather is arriving.

2.7 Questions for Review and Discussion

1 Having an identifiable brand for your organization or project helps people know about the work you are doing. Effective acronyms are short hand ways for people to refer to your work. SSWIM and WAS*IS are two acronyms that got some traction and broad acceptance. Think about what you consider the type of contribution you would like to make at the intersection of weather and society. What would you call your group? What acronym would you use? (www.acronymfinder.com)

2 This question is for those interested in weather and GIS, graphic information systems. This books emphasizes that there is no single "general public." The world is inhabited by heterogeneous groups of stakeholders. For this assignment, you will need to have access to Google EarthTM (available at http://www.google.com/earth/index.html), U.S. Census Data (available at http://www.census.gov/) and the Iowa Environmental Mesonet NWS Warning database (available at http://mesonet.agron.iastate.edu/cow/). Using the KML files made available through the Iowa Environmental Mesonet, download a recent tornado warning for a region of your choice. Import this tornado warning into Google EarthTM and answer the questions below: What time of year did the warning take place? What time of day? What day of the week? Were schools in session? Were most people at work? Was it rush hour traffic? Was it the middle of the night?

Considering that many counties sound the sirens for their entire area, assume that the counties under the tornado warning sound the sirens if the tornado warning intersects with that county. How many counties are under the warning? Using U.S. Census Data, how many people live in those counties? How many are minors (under 18 years of age)?

How many are seniors (over 65 years of age)? How many are minorities? How many live under the poverty level? How do these different affect the vulnerability of the people in this warning?

Look at the warning in Google EarthTM and use the place data and search tools to estimate the following answers (note that Google EarthTM place data is not always accurate, but for this exercise assume that it is).

Are there any schools under the warning? Are there any universities under the warning? Are there any airports under the warning? Are there utility companies under the warning? What other stakeholders can you find using the data available in Google EarthTM? Consider the types of concerns that these different sets of stakeholders would have during a tornado warning.

2.8   Using What You've Learned: Homework Assignment From the Chapter

1   People talk about the weather all the time. Weather affects decision making on a daily basis. However, weather is not the only factor that influences how people conduct their daily activities. For this assignment, listen to the conversations around you. How do the people in your life talk about the weather, such as your parents, siblings, family members, or friends? Why are they talking about the weather? Did the weather or the weather forecast change their plans, make their day better, or interrupt their routine? What sources of weather information do they mention? Do they mention why that is their preferred source? Do they watch local television stations? If so, do they have a favorite TV meteorologist? Do they have a weather app for their phone? Do they use an informal metric, environmental cues, aching bones, or other forecasting method for their weather forecasts, such as looking at the sky or using a landmark to measure wind speed or rainfall? Do they mention what they wish they knew about the upcoming weather or what they wish the weather would do?

2   Be aware of what people around you are saying and take note of the variety of conversations occurring and the many ways that weather plays a role in people's lives. Select two people who are close to you. Interview a family member and a friend or colleague.

Write a one-page summary of how weather and weather forecasts affect these two people. Be as specific as possible. The questions in the previous paragraph will be a good start.

References

Adamson, G.C.D. (2011) The languor of the hot weather: Everyday perspectives on weather and climate in colonial. *Bombay Journal of Historical Geography*, 38: 1819–1828.

Allen, S. (2016) Oklahoma City rolls out new tornado warning system. *Oklahoman*, March 10.

Boseley, E. (2015) Moderately cold temps more deadly than heat wave. *The Guardian*, May 24.

Burns, K. (2008) Hurricane Gustav prompts responders to evacuate pets. *Journal of the American Veterinary Medicine Association*, 233: 1201–1202.

Burton, I., Kates, R., and White, G.F. (1993) *The Environment as Hazard*. New York: Guilford Press.

Committee on Disaster Research in the Social Sciences: Future Challenges and Opportunities, National Research Council. (2006) *Facing Hazards and Disasters: Understanding Human Dimensions*. New York: National Academies Press.

Cutter, S. (2013) Preparing for the worst should be a nonpartisan issue. August 2013. http://www.thestate.com/2013/08/21/2931514/cutter-preparing-for-the-worst.html#storylink=cpy 21 (accessed July 23, 2017).

Demuth, J.L., Gruntfest, E., Morss, R.E., Drobot, S., and Lazo, J.K. (2007) Weather and Society * Integrated Studies (WAS*IS): Building a community for integrating meteorology and social science. *Bulletin of the American Meteorological Society*, 88: 1729–1737.

Donovan, A. and Oppenheimer, C. (2011) The 2010 Eyjafjallajökull eruption and the reconstruction of geography. *Geographical Journal*, 177: 4–11.

Fiebrich, C.A. (2009) History of surface weather observations in the United States. *Earth Science Reviews*, 93: 77–84.

Gladwin, C., Gladwin, H., and Peacock, W. (2001) Modeling hurricane evacuation decisions with ethnographic methods International *Journal of Mass Emergencies and Disasters*, 19: 117–143.

Gladwin, H., and Peacock, W. (1997) Warning and evacuation: A night for hard houses. In: Peacock *et al.* (eds), *Hurricane Andrew: Ethnicity, Gender, and the Sociology of Disasters*. London: Routledge, pp. 52–74.

Gottlieb, R., and Klockow, K. (2013) Discussion on the WAS* IS Facebook page. July.

Henson, R. (2010a) Blue skies and green screens: The history of weathercasting graphics. *Weatherwise*, 63: 32–37.

Henson, R. (2010b) *Weather on the Air: A History of Broadcast Meteorology*. New York: AMS Press.

Hoekstra, S.H. (2015) *Decisions Under Duress: Influences on Official Decision Making During Superstorm Sandy*. Ph.D. thesis in Coastal Resource Management, East Carolina University.

Hoekstra, S.H. (2012) *How K-12 school district officials made decisions during 2011 National Weather Service tornado warnings*. Masters thesis, Department of Geography and Environmental Sustainability, University of Oklahoma.

Janković, V., and Barboza, C.H. (2009) *Weather, local knowledge and everyday life: Issues in integrated climate studies*. MAST.

Kalkstein, L.S., Jamason, P.F., Greene J.S., Libby, J., and Robinson, L. (1996) The Philadelphia hot weather-health watch/warning system: Development and application. Summer 1995. *Bulletin of the American Meteorological Society*, 77: 1519–1528.

Klinenberg, E. (2003) *Heat Wave: A Social Autopsy of Disaster in Chicago*. Chicago: University of Chicago Press.

Larsen, J. (2006) Setting the record straight: More than 52,000 Europeans died from heat in summer 2003. Earth Policy Institute. http://www.earthpolicy.org/plan_b_updates/2006/update56 (accessed July 23, 2017).

Lazo, J.K., Morss, R.E., and Demuth, J.L. (2009) 300 billion served: Sources, perceptions, uses, and values of weather forecasts. *Bulletin of the American Meteorological Society*, 90: 785–798.

League, C.E., Philips, B., Bass, E.J., and Diaz, W. (2012) Tornado warning communication and emergency manager decision-making. Presentation at American Meteorological Society, January 24. http://ams.confex.com/ams/92Annual/flvgateway.cgi/id/20175?recordingid=20175 (accessed July 23, 2017).

League, C., Díaz, W., Philips, B., Bass, E.J., Kloesel, K., Gruntfest, E., and Gessner, A. (2010) Emergency manager decision making and tornado warning communication. *Meteorological Applications*, 17: 163–172.

League, C. (2009) *What Were They Thinking? Using YouTube to Observe Driver Behavior While Crossing Flooded Roads.* Unpublished masters thesis in Applied Geography, University of Colorado at Colorado Springs.

Liverman, D. (2011) Fostering advances in interdisciplinary climate change. Arthur M. Sackler Colloquia of the National Academy of Sciences. http://sackler.nasmediaonline.org/2011/climate_science/diana_liverman/diana_liverman.html (accessed July 23, 2017).

Maier, G., Grundstein, A., Jang, W., Li, C., Naeher, L.P., and Shepherd, M. (2014) Assessing the performance of a vulnerability index during oppressive heat across Georgia, United States. *Weather, Climate and Society*, 6: 253–263.

Masters, J. (2010) Over 15,000 likely dead in Russian heat wave: Asian monsoon floods kill hundreds more. Weatherunderground Wunderblog, August 9. www.wunderground.come/blog/JeffMasters/comment.html?entrynumber=1571 (accessed July 23, 2017).

McElroy, D. (2010) Russian heatwave kills 5,000 as fires rage out of control. *The Telegraph*, August 6. http://www.telegraph.co.uk/news/worldnews/europe/russia/7931206/Russian-heatwave-kills-5000-as-fires-rage-out-of-control.html (accessed July 23, 2017).

Mileti, D.S. (1999) *Disasters by Design: a Reassessment of Natural Hazards in the United States.* Philadelphia, PA: John Henry Press.

Montz, B., and Gruntfest, E. (2002) Flash flood mitigation: Recommendations for research and applications. Global Change. *Environmental Hazards*, 4: 15–22.

Moore, P. (2015) The birth of the weather forecast. *BBC Magazine*, April 30, 2015. http://www.bbc.com/news/magazine-32483678 (accessed July 23, 2017).

Morss, R.E., and Hayden, M.H. (2010) Storm surge and "certain death": Interviews with Texas coastal residents following Hurricane Ike. *Weather, Climate and Society*, 2: 174–189.

Nichols, A.C. (2012) *How university administrators made decisions during National Weather Service tornado warnings in the spring of 2011.* Masters thesis, Department of Geography and Environmental Sustainability, University of Oklahoma.

NOAA NWS. (2012) *NWS Strategic Plan, Building a Weather-Ready Nation* (also known as the NWS Roadmap).

Pielke, Jr., R., and Carbone, R. (2002) Weather impacts, forecasts, and policy: An integrated perspective. *Bulletin of the American Meteorological Society*, 83: 393–403.

Pielke, Jr, R. (1997) Asking the right questions: Atmospheric sciences research and societal needs. *Bulletin of the American Meteorological Society*, 78: 255–264.

Ruin, I., Creutin, J-D., Anquetin, S., and Lutoff, C. (2008) Human exposure to flash floods – Relation between flood parameters and human vulnerability during a storm of September 2002 in Southern France. *Journal of Hydrology*, 361: 199–213.

Semenza, J.C., Rubin, C.H., Falter, K.H., Selanikio, J.D., Flanders, W.D., Howe, H.L., and Wilhelm, J.L. (1996) Heat-related deaths during the July 1995 heat wave in Chicago. *New England Journal of Medicine*, 335: 84–90.

Sudan, R. (2008) Chilling Allahabad: Climate control and the production of Anglicized weather in early modern India. *Journal for Early Modern Cultural Studies*, Fall/Winter(8): 56–73.

Tobin, G. and Montz, B.E. (2015) *Evolving Approaches to Understanding Natural Hazards*. Cambridge: Cambridge Scholars Publishing.

Turner, R. (2009) Keeping meteorology masculine: The American Meteorological Society's response to TV 'Weather Girls' in the 1950s. In Jankovic, V., and Barboza, C. (eds.), *Weather, Local Knowledge and Everyday Life: Issues in Integrated Climate Studies*, pp. 147–158.

White, G.F. and Haas, J.E. (1975) *Assessment of Research on Natural Hazards*. Boston: MIT Press.

White, G.F., Kates, R.W., and Burton, I. (2001) Knowing better and losing even more: the use of knowledge in hazards management *Environmental Hazards*, 3: 81–92.

3

Social Science Partners and the Weather/Society Work They Do

The social sciences are a group of academic disciplines. Figure 3.1 shows the key social sciences addressing the problems of weather and society: anthropology, communication, economics, geography, psychology, and sociology. Other social science fields including political science, risk asessment, and science-technology studies are making contributions to the field as well.

This chapter provides a brief definition of six social sciences to demonstrate how they can contribute to the discussions about weather and society. In this overview, we will look at: 1) an example of how a social scientist from the particular discipline has approached a meteorologically relevant topic; 2) a list of other representative examples of work at the intersection of weather and society from that discipline's perspective. A more complete reference list is provided in the book's bibliography.

Every point of view is valuable for solving weather and society problems. The humanities have long contributed to our understanding of how people live with severe weather. The recent growth in social sciences work devoted to questions about weather has made it difficult to keep up with all new research and applications. Moreover, none of the perspectives is static; relationships, weather events, and experience influence what people want to know, what they need to know, and, when they need to know it. To solve problems at the intersection of weather and society, all partners have to be included throughout the problem-solving process. Just informing stakeholders of the outcomes or bringing them in at the end of a long process will not be effective. Partners need to talk and listen to each other from the start.

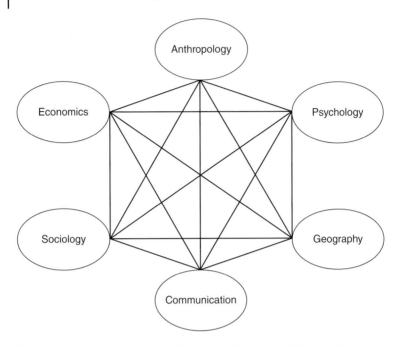

Figure 3.1 Key social science disciplines that address the problems at the intersection of weather and society.

Social science refers to disciplines whose primary objective is to help understand behavioral and social phenomena. "The social sciences are quite diverse but they all focus on some aspect of behavior and social life and on the institutions, technology, ideas and aesthetic creations emanating from social interactions" (Ellis, 1994:3). Like physical science, social science utilizes a range of methods including qualitative and quantitative techniques. Some methods are preferred by particular social science subfields and some methods are shared across the disciplines. For example, some of psychologist Dr. Michael Lindell's research on how households make decisions within time and space overlaps with geographical research.

The various disciplines in social science and physical science often don't communicate easily across disciplinary boundaries. There are practical reasons for this situation. For starters, our graduate education is structured so that a person usually studies within one disciplinary silo, learning as much as possible about the theories and methods of one field. Such a structure provides little incentive and few opportunities to look outside that discipline by reading each other's professional journals or walking across campus for lectures. Social scientists and physical scientists do not learn how to talk to each other. Also, individual personalities can also hinder progress toward more integrated approaches.

As part of professional preparation, physical and social scientists learn and use massive vocabularies and numerous acronyms with terms that are largely unknown outside the field. The first steps for moving outside of disciplinary boxes and developing collaborations involves time-consuming attempts to find common ground for discussing problems and terminology that translate jargon and acronyms so collaborators can share a common understanding.

Fruitful collaborations are growing thanks to new programs from funding agencies and inspiration from early career people who recognize the necessity to use multi-disciplinary approaches to solve complex problems. Evidence of cross fertilization of social science and physical science appears in social science research being published in physical science and engineering journals: (e.g., Laska and Morrow, 2006 in the *Marine Technology Society Journal*; Kates *et al.*, 2006 in the *Proceedings of the National Academy of Sciences*).

3.1   The Partners and How They Do Their Work

Figure 3.2 is a basic graphic showing how the people in various social science disciplines interact with the many partners who are concerned about weather and social science. These partners range from individuals who need to decide whether or not to bring a coat to work to decision makers for large school districts who face choices about closing school for snow, cold, or ice. Figure 3.2 shows how social sciences are related to a wide range of stakeholders including researchers, emergency managers, companies in the private sector, other government agencies, the media, and weather-forecasting software developers.

Figure 3.3 provides a simplified snapshot of differing weather-related questions that social scientists from these varying disciplines might ask. All these questions relate to human, organizational, or individual behavior and weather, but each discipline mostly considers

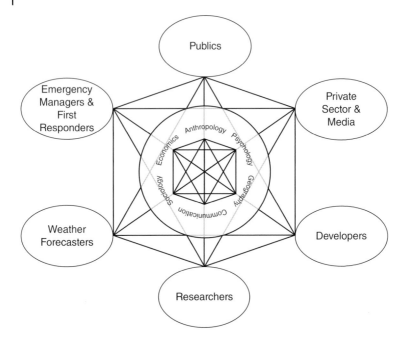

Figure 3.2 Social scientists from these disciplines solve weather impact problems for a range of decision-makers from individuals to emergency managers, and they work with other researchers and practitioners from the private and public sectors.

its own set of questions. The research at the intersection of social science and weather is in its infancy.

As of 2017, the quality and quantity of social science studies integrated with meteorology are rising in the United States and elsewhere. The German Weather Service completed surveys of how weather warnings are communicated to professional end users in the emergency community and how the warnings are converted into mitigation measures. Online questionnaires were completed by 161 members of emergency services including fire fighters, police officers and civil servants (Kox *et al.*, 2015).

Generalizing from these small studies and extrapolating to entire populations is difficult. Limited research findings from a few social science research studies haven't clarified exactly to what degree social science results relate to how people prepare for hurricanes or tornadoes.

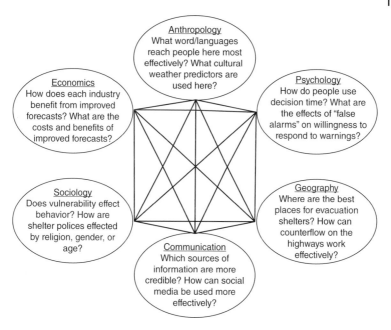

Figure 3.3 Examples of questions that social scientists ask at the intersection of weather and society.

Are the coping strategies for hurricane warnings and response fundamentally different than the strategies used in tornado situations? Weather events are quite different in terms of warning time until impacts and the types of impacts. Do residents in Columbus, Ohio respond the same way to tornado warnings as residents in Birmingham, Alabama or are there local geographic and meteorological elements leading to localized perceptions and responses?

Because the field is young and funding has been limited, there is much that social scientists do not know about how and why people behave when faced with severe weather and warnings. More social science research and the application of its findings to operations and policy should reduce losses from severe weather and allow for better decision making. The social science findings are quite limited. As of 2008, less than 1% of NOAA's research budget was allocated to social science (NOAA, 2009).

People get their weather information from different sources including broadcast meteorologists, emergency managers, and/or the NWS forecasters, making it important to consider the wide range of decision makers in any discussion of weather and society. Social science research related to weather shows that there is "no one size fits all" approach for changing how people behave when faced with severe weather or warnings of severe weather (Hayden *et al.*, 2007). Social scientists tackle a wide variety of weather-related questions, and the weather/society intersection involves a large number of users, stakeholders, and partners. Multi-year multi-faceted research projects, more than the current one-or two-shot case studies, are needed to understand how weather fits into decision-making contexts (Spinney and Gruntfest, 2012).

Most influential research regarding weather is conducted by social scientists from interdisciplinary teams where, for example, communication experts work with meteorologists and sociologists. Often more than one discipline is represented. The research team at the National Center for Atmospheric Research (NCAR) Societal Impacts Group has included people with backgrounds in anthropology, economics, meteorology and communication (e.g., Morss *et al.*, 2008; Morss *et al.*, 2005). Publications by interdisciplinary teams include Trumbo (2012); Morss *et al.* (2005); Schumacher *et al.* (2010); Lazo *et al.* (2009).

Many people working on weather issues still define themselves by disciplinary boundaries. Most academic and even federal institutions are still structured in disciplinary silos and most people are still hired along disciplinary lines. There is great depth within each discipline, and each discipline has experts and specialists. Understanding differences and commonalities is critical to getting the right input for any particular problem at the intersection of weather and society.

3.2 Anthropology

Anthropology means "the study of humankind" (Ellis, 1994:4). Many anthropologists study the diverse cultures of humans. Cultural anthropologists specialize in studying the diversity of human cultures and social customs as exhibited by people in thousands of human societies throughout the world (Ellis, 1994:4).

Anthropology deals with the origin, development, and nature of humans and their cultures. It differs from other social sciences in its "comparative" approach (Smith and Fisher, 1970:5). Anthropologists often collect data through the use of what is termed "fieldwork" in which the anthropologist spends an extended period of time living with the people being studied.

With regard to weather, anthropologists might ask:

- What *cultural* factors are most influential in identifying who will and who will not follow official orders to evacuate?
- What *cultural* characteristics affect the use of public shelters?
- Do local residents perceive "severe" weather in the same ways as public officials or meteorologists?

A few anthropologists study people and cultures at the intersection of weather and society, and one tool these anthropologists use is to retell or convey the stories told to them by their informants (Peterson and Broad, 2008). They use the term "narrative" to refer to stories and myths shared by field consultants. These narratives can suggest ideas and norms, actions or perspectives that describe how people's lives intersect with weather.

A growing body of anthropological research on climate change reflects the biases, interests and directions of previous climate-related studies (i.e., weather and seasonal climate variability), as well as how social scientists have addressed general environment-human and environment-society interactions (Oliver-Smith, 1996; McNeeley and Lazrus, 2014).

Anthropological research shows that arctic sea ice is becoming thinner, freezing later and thawing earlier, wind becoming more unpredictable, changing usual direction, and more frequently reaching storm strength, permafrost thawing, and animals changing migration patterns and declining in numbers (Ford *et al.*, 2010:179).

Dr. Karen Pennesi, Professor of Anthropology at University of Western Ontario, has studied how Brazilian farmers used weather forecasts from official government agencies and from weather "prophets" to make decisions about planting their crops (Pennesi, 2011, 2007). In 2012, she and her colleagues published their study on how the variability in weather patterns is changing the ways of life in Nunavut, one of Canada's Arctic territories. The Nunavut people reported to the researchers that their travel and land-based activities are becoming more risky (Spinney and Pennesi, 2012).

Dr. Pennesi and her colleagues found that the Nunavut people "lack sufficiently reliable and useful information on which to base their decisions" (2012:897). Their research methods consisted of participant observation and semi-structured interviews with indigenous and non-indigenous long-term residents in summer 2009. They studied how the residents of Iqaluit, the capital of Nunavut, acquire, perceive and use both local and scientific weather knowledge.

One of the research team members who lived in Iqaluit for two months took trips with the local residents and was able to "experience, document, and discuss hazards as they were encountered. The researchers developed an understanding of local practices, attitudes and knowledge that goes beyond what was revealed in interviews" (Pennesi *et al.*, 2012:904). They found that experienced hunters were perceived to be a reliable source of weather-related information, "while scientific weather knowledge is not as accessible or informative as it could be" (2012:897).

Pennesi *et al.* (2012) asked seven sets of questions related to the following topics: 1) length of residency; 2) what is considered hazardous weather?; 3) what are the impacts of hazardous weather?; 4) have you ever experienced damage to your property due to storms of other bad weather?; 5) how do you receive weather information?; 6) what social networks do you rely upon? (Are there certain people you would ask for advice before going out on a trip?); and 7) what do you do to prepare for bad weather when a storm warning is issued?

They found that information on hazardous weather and environmental conditions is shared among residents who are active on the land, have access to various social networks, and who speak Inuktitut. This weather information is largely inaccessible to Anglophone and Francophone communities or to newcomers who want to get out on the land. They found that Iqaluit residents often use their own local knowledge to read weather conditions and assess likely hazards. Environmental phenomena such as clouds, sky, behavior of wildlife, direction of the tide flow, snow drifts, and direction of currents were listed as local forecasting indicators (Pennesi *et al.*, 2012:908).

Figure 3.4 shows how the researchers coded the data from their interviews for their analysis. Based on their research, Pennesi *et al.* made five main recommendations to the government policy makers: 1) form a weather hazards impact advisory group; 2) develop more formal and consistent documentation and dissemination of local knowledge; 3) provide hazard maps and marking of hazardous locations; 4) integrate weather-related information sources; and 5) share information across language groups.

The study calls for more research that looks at how factors such as ethnicity, language, age, length of residence, and the nature and extent

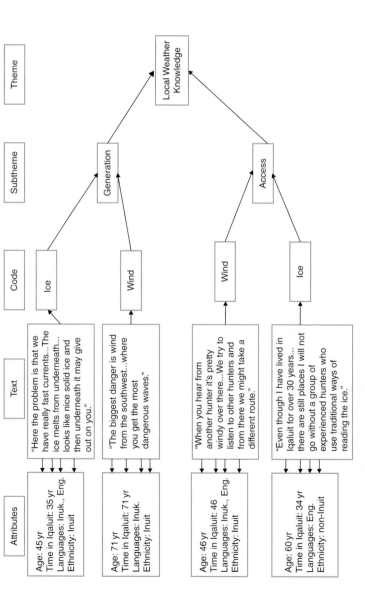

Figure 3.4 Integrating local and scientific weather knowledge as a strategy for adaptation to climate change in the Arctic. *Source: Pennesi et al., 2012:907. Reproduced with permission from Springer.*

Attributes	Text	Code	Subtheme	Theme

Age: 45 yr
Time in Iqaluit: 35 yr
Languages: Inuk., Eng.
Ethnicity: Inuit

"Here the problem is that we have really fast currents…The ice melts from underneath… looks like nice solid ice and then underneath it may give out on you."

Age: 71 yr
Time in Iqaluit: 71 yr
Languages: Inuk.
Ethnicity: Inuit

"The biggest danger is wind from the southwest…where you get the most dangerous waves."

Age: 46 yr
Time in Iqaluit: 46
Languages: Inuk., Eng.
Ethnicity: Inuit

"When you hear from another hunter it's pretty windy over there…We try to listen to other hunters and from there we might take a different route."

Age: 60 yr
Time in Iqaluit: 34 yr
Languages: Eng.
Ethnicity: non-Inuit

"Even though I have lived in Iqaluit for over 30 years… there are still places I will not go without a group of experienced hunters who use traditional ways of reading the ice."

Ice

Wind

Wind

Ice

Generation

Access

Local Weather Knowledge

of land-based activities influence perceptions and uses of weather information. They also call for increased interaction between local communities and the Meteorological Services of Canada, emphasizing that any efforts must be developed in collaboration with communities or local policy makers. In earlier research, Ford and colleagues (2010:188) found that recommendations need to be identified and developed in collaboration with communities or policy makers if they are going to have the required legitimacy and integration of Inuit knowledge that are essential to decision making in the new North. This collaboration will also help explain how adaptation to changes in weather links into the policy process.

Another interdisciplinary team of an atmospheric scientist, a computer scientist and an anthropologist evaluated whether meteorological data related to more extreme weather matched public perception that local conditions were changing in the Arctic. The native people were noticing that conditions were no longer as reliable for as long as they had been. The researchers documented that the meteorological data backed up the local peoples' perception that weather conditions did not "persist" as long as they had in the past (Weatherhead *et al.*, 2010).

Other anthropologists who study weather include Strauss and Orlove (2003), Oliver-Smith (Oliver-Smith and Hoffman, 1999; Oliver-Smith, 1996) and Gladwin (Gladwin *et al.*, 2001; Gladwin *et al.*, 2007; Gladwin and Peacock, 1997).

3.3 Communication

Effective human communication is at the center of personal, social, and natural harmony (King, 1989:1). "The primary function of communication is to establish relationships between human beings and to provide a means for monitoring and maintaining or terminating such relationships" (King, 1989:2). The process of communication, commonly defined as the sharing of symbols over distances in space and time, encompasses a wide range of topics and contexts ranging from face-to-face conversation to speeches to mass media and social media (King, 1989).

At the weather/society interface, communication experts help develop effective methods of getting the messages of weather products and services to various stakeholders; that means all of us who are impacted by weather. They understand best practices in presenting and communicating uncertainty in scientific assessments related to weather. Weather-relevant questions that communication researchers might ask include:

- How are social media (e.g., Facebook, Twitter) effectively used before, during, and following hurricanes evacuations? and
- What communication sources do people trust most?

Dr. Julie Demuth's degree is in technical journalism. She also has a master's degree in atmospheric science from Colorado State University. She is a research scientist at NCAR in the Societal Impacts Program. On several research projects, Dr. Demuth teamed with Dr. Rebecca Morss, a meteorologist; sociologist Dr. Betty Morrow; and economist Dr. Jeffrey Lazo, an economist. A recent study of the hurricane warning and response process is described here.

Hurricane forecasts have improved significantly in recent years. "The average error in hurricane track forecasts has been reduced by 60 percent since 1990 (NOAA, 2011) and there is a growing social science research literature directed at understanding how vulnerable people make preparedness and evacuation decisions" (Lazo *et al.*, 2014).

Dr. Demuth and her colleagues examined the starting point of the warning and response process using a mixed-method empirical approach. They relied mostly on in-depth, semi-structured interviews with members of three groups.

They asked interviewees' in the Greater Miami, Florida area about job roles and partnerships, their sources, uses, creation and communication of information, their audiences, and their views on the hurricane forecast and warning process (Demuth *et al.*, 2012:1133). Demuth *et al.* studied how NWS forecasters at the National Hurricane Center and local weather forecast offices, local emergency managers, and local television and radio media create and convey hurricane risk information. They also visited and observed members of these groups from Greater Miami to learn about their roles, goals, and interactions, and to identify strengths and challenges in how they communicate hurricane threats with each other and with the public.

Some forecasters reported that the information they provide is not used as they intended. Media personnel wanted streamlined information from the NWS and emergency managers emphasizing the timing of hazards and recommended responses and protective actions. Emergency managers asked for forecast uncertainty information to help them plan for different scenarios. The research team also developed "mock" scenarios of a hurricane four days away from striking Miami, Florida. Interviews were recorded and the data were analyzed.

Demuth *et al.*'s research goal is to improve how partners in the warning process interact with each other and how they generate and convey information. Each partner sees its role and the other partners through a particular lens that the researchers try to expand by highlighting differing points of view in the hurricane communication process. They studied how information about forecasts, warnings, and recommended protective actions is created and communicated when a hurricane threatens. Others have studied how people interpret forecasts that include probability of precipitation (Gigerenzer *et al.*, 2005).

Demuth *et al.* recommend four steps for these warning communication partners: 1) build understanding of each other's needs and constraints; 2) ensure formalized, yet flexible, mechanisms for exchanging critical information; 3) integrate social science knowledge to design and test messages with intended audiences; and 4) evaluate, test and improve the NWS hurricane–related products in collaboration with social scientists (2012:1133). The authors use quotes from the interviews to emphasize their points. One forecaster said, "The media always want the data sooner, quicker, they want insights. Can you whisper it in my ear type of thing... They typically want more details, more specifics. They want more content. We tend to be very conservative, maybe rightfully so, on content, given the uncertainty. We don't want to hype things up, so they often want more detail, more specifics, prettier pictures, things they can show on air, maybe things that are not appropriate to show on air" (Demuth *et al.*, 2012:1140).

Their work demonstrates a great deal of variation in what people think a particular forecast means. The researchers asked people: What do you think it means when the forecast states there is a 60% chance of rain for tomorrow? More than 80% did not understand what the forecast meant.

Demuth and her colleagues also completed a study in the United States asking about forecast confidence based on different lead times. Up until two days out, the public has high confidence in the forecasts. Based on their sample between 40% and 50% of the public has high confidence in one-day forecasts for temperature and chance and amount of precipitation. These studies provide baseline data to be used for comparison as new weather products and communication methods are developed and disseminated.

In a fourth example, Demuth *et al.* examined the ways people wanted to receive their weather information. There are many discussions about whether or not people understand probabilistic or deterministic forecasts. Deterministic forecasts simply say whether it will rain—yes, or no. Probabilistic forecasts offer a range of likelihoods. Their research shows that people already expect television forecasts to include uncertainty or likelihood rather than ironclad forecasts.

Other communication scholars working on weather studies include Dr. Craig Trumbo, communication professor at Colorado State University (Trumbo *et al.*, 2012), Dr. Susan Jasko, communication professor at California University of Pennsylvania and Dr. Gina Eosco, a research scientist with the NOAA (Eosco, 2015, Dixon *et al.*, 2015 and Freberg *et al.*, 2013).

3.4   Economics

Economists study all aspects of economic processes, from individual and family financial well-being, to the financial affairs of institutions, states, and nations (Ellis, 1994:5). Economists study the production, distribution, and consumption of goods and services. Economists, who are interested in weather-related problems, analyze the costs of weather-related disruption and destruction. Economists often establish net benefits of programs for budgetary justification and program evaluation, for understanding incentives of participants and stakeholders in organizational and economic processes relevant to NOAA missions, and for studying "human behavior as a relation between scarce means having alternative uses" (Lazo, 2010).

Perhaps because they rely on quantitative metrics that atmospheric scientists are most comfortable with, economics has a longer track record in collaboration with physical scientists than the other social sciences. In industrialized countries, about 70% of firms are exposed to changes in everyday weather in a wide range of economic sectors, including agriculture, tourism, food, beverage, transportation, and construction.

To establish net benefits of programs for budgetary justification and program evaluation, economists might ask:

- When preparing for a hurricane evacuation, officials are considering making major highways all one-way to allow cars to only head away from the shore (counter flow)—what are the economic benefits and costs of counter-flow operations?

- How much do hurricane evacuations cost?
- What do major chain stores do in anticipation of a major hurricane like Superstorm Sandy (Souza, 2012)?
- What are the economic impacts of evacuating coastline motels during a holiday weekend when hurricane warnings are issued and the storm ends up being less severe than expected or misses that part of the coastline?

Economists use monetary values to calculate the impacts of severe weather. They add up losses to crops, livestock, timber, and homes. The costs can be considered direct losses such as washed away businesses, or indirect losses such as lost productivity or impacts on local businesses from roads being washed out for long periods of time. Social/psychological and ecological/ecosystem impacts normally are not counted as economic impacts.

Dr. Jeffrey Lazo, an economist at NCAR, has conducted economic studies of the value of weather and weather information to the U.S economy. He also suggests how meteorologists can effectively incorporate economics into their understanding of the value of weather, "… There is a wealth of knowledge from economics that is being essentially ignored by those using the cost–loss model for assessment of the value of weather forecasts" (Lazo, 2010:171).

Two recent economics studies by Lazo and his colleagues looked at: 1) the value of weather information to the U.S. economy; and 2) the value of weather forecasts to the U.S. public. Their first study showed that routine weather events such as rain and cooler-than-average days can add up to an annual economic impact of as much as $485 billion in the United States (UCAR, 2011). Like the communication studies cited above, this study is the first attempt to synthesize a very complex issue. The researchers did not evaluate the possible impacts of climate change, which is expected to lead to more flooding, heat waves, and other costly weather events. Still, the study concludes that the influence of routine weather variations on the economy is as much as 3.4% of U.S. gross domestic product.

The study found that finance, manufacturing, agriculture, and every other sector of the economy is sensitive to changes in the weather. The impacts can be felt in every state.

It's clear that our economy isn't weatherproof... This is the first study to apply quantitative economic analysis to estimate the weather sensitivity of the entire U.S economy. The research may help policymakers determine whether it is worthwhile to invest in enhanced forecasts and other strategies that could better protect economic activity from weather impacts (UCAR, 2011).

A nationwide survey indicates that the U.S. public obtains several hundred billion forecasts each year generating $31.5 billion in benefits compared to costs of $5.1 billion (UCAR, 2009 press release; http://www2.ucar.edu/news/845/300-billion-weather-forecasts-used-americans-annually-survey-finds). Lazo and his colleagues found that nearly 9 out of 10 adult Americans obtain weather forecasts more than three times a day, on average. "The value Americans place on these forecasts appears to be far more than the nation spends on public and private weather services." Using an internet survey of 1520 respondents, 1465 said that they used weather forecasts. Lazo and his colleagues asked, on a per-household basis, how much would respondents value each weather forecast? The average value was 10.5 cents per forecast. "This equates to an annual value of $31.5 billion." The average respondent gets weather forecasts 115 times per month. Nearly 226 million U.S. adults access more than 300 billion forecasts per year. The authors found that many people use forecasts for planning specific activities, such as vacations, and routine daily activities. They use the forecasts to help them make many decisions including what to wear and how to get to work or school. The study revealed that peak periods for accessing forecasts are early morning, early evening and late evening (Lazo *et al.*, 2009).

Some approaches economists have developed relevant to the value of weather forecasting include the following:

- A strong theoretical and applied understanding of public goods that would include weather forecasts that 1) helps explain why we do not have information on the value of forecasts and 2) provides theoretical justification for the public provision of weather information (Craft, 2010);
- A long history of theory, methods, and applications looking at decision making under risk and uncertainty, which is applicable to the

use of weather forecasts in decision making (Friedman and Savage, 1948);

- Nonmarket valuation methods for determining the value of public goods that is just beginning to be applied to the topic of weather forecasting (Lazo *et al.*, 2011);
- Applied methods for estimating the benefits and costs of specific changes in weather forecast systems (Sutter and Erickson, 2010); and
- A range of other issues relevant to weather forecasting, such as risk preferences, information asymmetry, discounting, value of statistical life, and applications to specific industries (Lazo, 2010:171).

Lazo *et al.* asked the question: What is the value of weather information to the U.S. economy?

> Weather directly and indirectly affects production and consumption decision making in every economic sector of the U.S at all temporal and spatial scales. From very local, short-term decisions about whether or not to pour concrete on a construction project to broader decisions of when to plant or harvest a field, to the costs of rerouting an airplane around severe weather, to predicting peak demand electricity generation in response to extreme heat, or to forecasting early season snow for a bumper ski season in Colorado, drought in the Midwest, or wind-fueled wildfires in California, weather can have positive or negative effects on economic activity (Lazo et al., 2011:1).

Lazo *et al.* (2011) consider how variability in weather patterns affect manufacturing, agriculture, transportation, recreation, utilities and other key elements of the U.S. economy. Private sector companies including utilities, weather forecasters, ski resorts, and many others collect their own sets of data but do not make their results known outside of the company. Normal weather variation, excluding extreme events and disasters, cost the nation an average of $485 billion a year or 3.4% of the 2008 national economy. Their study showed that manufacturing, agriculture, and every other sector of the economy is sensitive to changes in temperature and precipitation and that impacts can be felt in every U.S. state.

The authors used a nonlinear regression analysis, a statistical technique for comparing multiple variables. They examined 70 years of weather records, from 1931 to 2000 focusing on temperature (heating degree days and cooling degree days), total precipitation and deviations from average precipitation. They then divided the private economy into 11 sectors. They examined the sensitivity of these sectors to weather variability using 24 years of state-level economic data. This is the period for which detailed state-level data were available and consistent for major economic sectors. In their analysis, the authors focused on the 48 contiguous states, excluding Alaska and Hawaii. They explained these two states were outliers and represented only 0.6% of the total U.S. Gross Domestic Product (GDP).

The results showed the complex influence of weather. For example, a prolonged dry spell is terrible for crops but good for construction projects. A snowstorm might disrupt air travel and drive up heating costs but it usually boosts attendance at ski resorts. The report also concluded that the economy of every state is sensitive to the weather. New York was most sensitive with a 13.5% impact on the gross state product. Tennessee was least sensitive with a 2.5% impact on the gross state product. The state-level findings were more subject to error than national findings, but overall the authors noted that when aggregated across all 11 sectors, "no one part of the country appears significantly more weather sensitive than another region in relative terms" (http://journals.ametsoc.org/doi/pdf/10.1175/2011BAMS2928.1).

The United States as a whole is more resilient than the states. Bad weather in one area is usually compensated with good weather in another. Economic production can shift from one region to another. There were weather-related losses every year during the 70 years between 1931 and 2000. The greatest losses were in 1969, and 1939 had the least. When the authors applied this to the 2008 national GDP, the study indicated that routine weather events such as rain and cooler-than-average days can add up to an annual economic impact of as much as $485 billion.

In other economic research, a 2014 economic study in Switzerland examined the economic value of meteorological services to the country's airlines by studying weather forecasts for the Zurich airport and for the country's transportation system (Frei, 2010; Von Gruenigen *et al.*, 2014). Researchers found a cumulative economic benefit for all domestic airlines at the two main Swiss airports of between 13 to 21 million USD per year. Using qualitative interviews to complement their quantitative economic analyses, they found that organizational factors are difficult to separate from meteorological forecasting information in terms of the differences they made to decision making at airports. Von Gruenigen *et al.* suggest that their economic methods can be applied to other studies of the economic benefits of meteorological services (2014:271). Their methods begin with the decision-making context within companies.

Other economists who work at the intersection of weather and society include David Letson, economics professor at University of Miami, who studies the value of hurricane forecasts and mitigation (Letson *et al.*, 2007). Some researchers evaluate weather and consumer behavior (Murray *et al.*, 2010). A classic work by Katz and Murphy is the book *Economic Value of Weather and Climate Forecasts* from 1997 (Katz and Murphy, 1997). Simmons, economics professor at Austin College, often collaborates with economist Dan Sutter professor at Troy University on numerous quantitative economic weather research projects related to the economic value of lead time for tornadoes and false alarms (2011, 2009, 2008, 2007, and 2005). Predicting how weather will impact crop prices or other commodities is a growing field in finance and economics (Pollard *et al.*, 2008).

3.5 Geography

Geography focuses on "how political and economic processes affect and are affected by features of the earth at or near the earth's surface" (Ellis, 1994:5). Geographers study people and the environment in which they live. *Geography* is the study of the earth and its features, its inhabitants, and its phenomena. Most geographers are trained in human and physical geography. Not all geographers are social scientists but even physical geographers, who are experts in landforms and climate, have some exposure to the social science elements of geography. Most geographers study weather and climate as part of their undergraduate coursework. Geographers usually identify as a physical or a social scientist. Human geography focuses largely on the built environment and how space is created, viewed and managed by humans, as well as the influences that humans have on the space where they live and work. Since the 1930s, there has been a tradition of geographers working in interdisciplinary research teams covering a wide range of natural hazards including meteorological ones including flooding, hurricanes, and debris flows. Gilbert F. White and his generations of students addressed extreme weather as part of their research.

Geographers' main concern is "where," and they use geospatial techniques such as geographic information systems (GIS), remote sensing, statistics, and cartographic mapping. They focus on temporal and spatial scales. Often geographers organize their work by scale. They consider the individual, the family, the community, the province, the region, the nation, and then the global scale. Because of their study of both human and physical aspects, many geographers are comfortable with integrated approaches to weather-related issues.

Geographer Dr. Isabelle Ruin and her colleagues at LTHE (Laboratoire d'étude des Transferts en Hydrologie et Environnement) in Grenoble, France focus on post-flash flood field studies. Warnings, if they exist at all, are very short fused. Space and time scales involved either in hydro-meteorology or human behavior and social organizations sciences are of crucial importance. Interdisciplinary collaboration is particularly important here because those involved with such events, including hydrologists, meteorologists, road users, emergency managers, and civil security services, all have different time and space frameworks that they use for decision making, forecasting, warnings, and research. The work of Dr. Ruin and her team examines the consequences of different response timescales for a river, the public, and forecasters.

They studied catastrophic flash floods in southern France from 2002 and in 2010 that killed a total of 50 people. Motivated by the need to reduce flood losses, Ruin *et al.* studied the choices and constraints of travel during flash flood events, and they identified the links between risk perception and spatial practices. They used quantitative and qualitative methods that included interviews, questionnaires and cognitive mapping surveys. They demonstrated that people do not correctly locate flood risks on roads, a serious problem for drivers on roads adjacent to small catchments with short response times.

They also found that drivers underestimate the depth of water that is dangerous for their cars. Ruin *et al.*'s research evaluates how two major daily priorities influence travel choices in crisis periods (Ruin *et al.*, 2007). People reported that they "had to" go to work. Their research showed that middle-aged workers in cars are most vulnerable to flash floods. The workers do not have the discretion to cancel their trip, or change routes to avoid dangerous low water crossings, even if they perceive the risk posed by a flash flood as being high. The workers believe the risk of losing their job is greater than the risk of driving in dangerous conditions. Parents reported that they would pick up their children from school as soon as warnings are issued even if officials forbid it. People who were retired or who had more flexible schedules reported that they were more willing to cancel their travels when weather watches were issued. The researchers integrated physical science findings from meteorology and hydrology into their social and demographic data.

Dr. Ruin and her colleagues compare crisis behavior with normal every day travel behavior. Ruin *et al.* (2008) used large census datasets related to urban mobility to find baseline data. Then they studied actual drivers' behaviors. The observation sites were selected based on a high flash flood hazard potential. Creating an inventory of these dangerous road points was a long process and required collaboration with road departments and law enforcement agencies.

Ruin *et al.*'s research is theoretical and applied. One of its main objectives is to make recommendations for how decision makers, from official to individual, can reduce flash flood losses. When Dr. Ruin and her colleagues study the choices people make and constraints of travel during flash flood events, they also examine links between risk perception and spatial practices. The combined analysis of interviews, questionnaires, and cognitive mapping surveys show that people do not correctly locate flood risks on roads.

Dr. Ruin and her colleagues combine the analysis of physical and human responses to Mediterranean storms. They use the results of hydrometeorological simulations, and they implement quantitative and qualitative research tools for interviews of flood victims, business owners, political leaders, emergency managers and others.

After 2010 flash floods in the Var region of France, Dr. Ruin and her colleagues conducted post-disaster fieldwork designed to learn from that disaster and to reduce future flash flood vulnerability. The work involved interdisciplinary teams that collected physical and social science data. In addition to learning how high the water was and how much rain fell, researchers asked residents and tourists if they heard any official warnings, what the environmental cues were, and how they changed their "normal" behavior during the flood. The research team also talked with flood survivors and officials to learn when they realized that there was a crisis that required changes in behavior. The researchers were particularly interested in stories they heard about small groups of people in stores or offices that informally organized, without any official warnings or directions, to get to higher ground.

Many lives were saved by these sensible actions. Ruin *et al.* examined what set of conditions prompted people to take such measures.

Dr. Ruin and her colleagues have strong interdisciplinary partnerships that go beyond common academic boundaries, and they incorporate the views of business owners, political leaders, and emergency managers. They try to understand how individuals and households adapt their daily routine (travel patterns and schedule) to brutal environmental perturbations. She and her colleagues are: 1) identifying cognitive and situational factors as well as thresholds that help people to switch from normal daily activities to adapted crisis response; 2) understanding the spatio-temporal interaction between individuals' decision-making processes and the dynamic of both the phenomena and the social response; and 3) studying socio-economical and cultural profiles of people showing resilient and vulnerable practices (Ruin *et al.*, 2014).

Figure 3.5 shows how space and time scales for the 2002 flash flood event can be integrated to understand when people died in the flood,

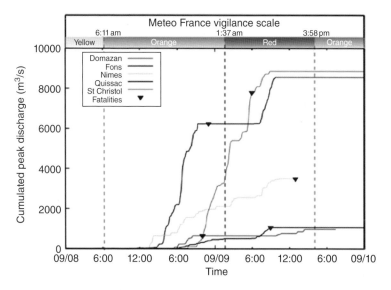

Figure 3.5 Hydrological, meteorological, and sociological data from one flood. Integrating the datasets to understand a flood in terms of the where and when of stream response, people's behavior, and precipitation patterns has great potential for increasing understanding of what happens during floods as part of research aimed at mitigating their impact. *Source:* Ruin *et al.* (2008). Reproduced with permission from Elsevier and I. Ruin.

what the hydrological conditions were and the type of warning that was in effect at the time. Meteo France uses a vigilance scale as their warning system where conditions are expected to deteriorate from yellow, to orange to red. Figure 3.5 shows that for the five flood fatalities, three occurred during the red warning and two occurred during the orange phase. The research that Dr. Ruin and her colleagues are doing brings hydrologists, meteorologists, and social scientists together on the same graphics, and their work is integrated across disciplines and time and space scales as shown in Figures 3.6 and 3.7 (Ruin *et al.*, 2014; Terti *et al.*, 2015). It has strong potential to transform hazards research and deserves a lot of attention from scholars and practitioners in many disciplines.

Dr. Bob Brinkman, a geographer, is the Director of Sustainability Studies and the Director of Sustainability Research at the National

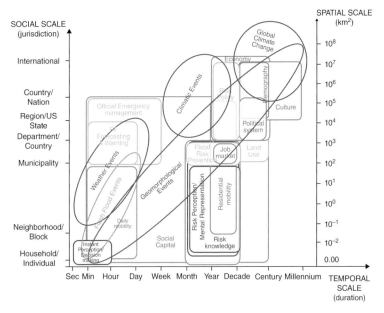

Figure 3.6 All of these systems interact in time and space to affect warnings for flash floods and responses to those warnings. *Source:* Ruin *et al.*, 2014. Reproduced with permission from I. Ruin and G. Terti. Flash flooding and the Global Environmental Change perspective: Toward a scaling approach for Disaster Reduction. Paper presented at the World Weather Open Science Conference, Montreal, Canada, August 16-21.

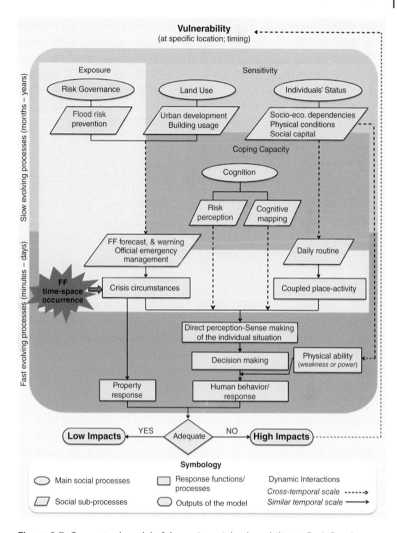

Figure 3.7 Conceptual model of dynamic social vulnerability to flash flood events
Source: Terti *et al.*, 2015:1482. Reproduced with permission from Springer.

Center for Suburban Studies at Hofstra University where he is also a Professor in the Department of Geology, Environment, and Sustainability. One of his research areas is the great variation in how a community prepares for hurricanes. Even though the National Hurricane Center in Miami serves all counties and cities

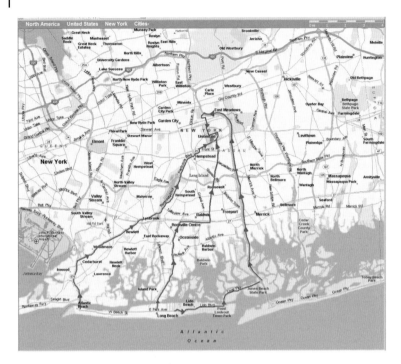

Figure 3.8 Nassau County hurricane evacuation map -Source: www. nassaucountyny.gov/2931/Hurricane-Evacuation-Routes. *Source:* Office of Emergency Management, Nassau County, Long Island, New York.

in the United States vulnerable to hurricanes, Brinkman points out that two counties that are both very vulnerable to hurricanes—the Hillsborough County Tampa Bay area in Florida and New York's Nassau County—have different approaches to planning (Brinkman, 2013).

The Hillsborough County Tampa Bay maps show expected storm surges by street level (Figure 3.9). The Nassau County one does not (Figure 3.8). The Tampa Bay maps (available on the website) also have extensive information about shelters and much more practical information including what to do before, during, and after a storm. The maps also provide information about insurance and solid links to a variety of resources. Brinkman thinks the Tampa Bay guides are a "gold standard." He has an active blog at http://bobbrinkmann. blogspot.com.

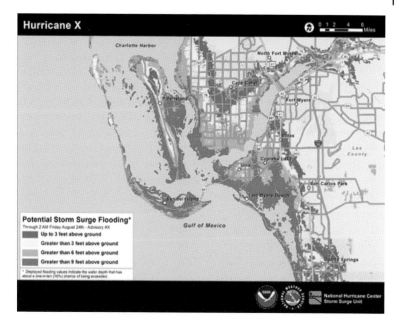

Figure 3.9 The National Weather Services created a prototype using Tampa-St. Petersburg, Florida for its new storm surge forecasting map using data collected when Hurricane Charlie struck Punta Gorda in 2004. http://www.tbrpc.org/tampabaydisaster/hurricane_guide2012/HG_2012.html *Source:* National Hurricane Center, National Weather Services, NOAA.

Other geographers who work at the intersection of weather and society include Dr. Jean Andrey who is a Professor of Geography at University of Waterloo in Canada (2010). Andrey studies how weather affects traffic and accident patterns. Dr. Susan Cutter (Cutter, 2013; Cutter *et al.*, 2007; Cutter and Smith, 2009) works extensively on hazard vulnerability and has created widely adopted spatial indices for the United States.

3.6 Psychology

Psychology is a science that tries to understand behavior and mental process and to apply that understanding in the service of human welfare (Bernstein *et al.*, 2011:4). Psychologists study and try to predict, improve, or explain how people behave and the mental processes that account for their behavior (Bernstein *et al.*, 2011:4). Psychology is the study of the mind, both conscious and unconscious, and how it influences behavior. Cognitive psychologists study mental processes underlying judgment, decision making, problem solving, and other aspects of human thought or cognition (Bernstein *et al.*, 2011:4). Related to weather, psychologists explore such concepts as perception, cognition, emotion, motivation, and interpersonal relationships related to behavior (Bernstein *et al.*, 2011:4).

Psychologists have taken many different approaches to weather questions. Dr. Michael Lindell is a social psychologist who has researched natural hazards mitigation since the 1970s. One of his research areas is trying to understand how households make decisions about their risks when evacuating or not evacuating in response to hurricane warnings from the National Hurricane Center.

He and his colleagues confronted the conventional wisdom that the decision to evacuate is simple. When a hurricane warning is issued, there are many considerations that make an evacuation decision much more than a go/no go decision. The decision of whether or not to evacuate is much more complex than just hearing a warning and then leaving (Lindell *et al.*, 2007). For many people, the evacuation time is defined by many factors including the time needed to prepare to leave from work, travel from work to home, gather all persons who would evacuate, pack items needed while gone, protect property from storm damage, shut off utilities, secure the home, and reach the main evacuation route (Lindell *et al.*, 2005:172).

Lindell and his colleagues' work is the most comprehensive on evacuation time estimates from the public official and household points of view. Lindell *et al.* (2005:172) found out from people who were asked to evacuate how much time people spent preparing to evacuate following a hurricane warning. Preparation time estimates ranged from approximately 60 to 450 minutes, with a mean of 229.9 and a standard deviation of 85.2 minutes. Lindell's team calculated that the elapsed time between the respondents' decisions to evacuate and the time they arrived at a major evacuation route was 196.2 minutes or just over three hours. However, 25 percent of the evacuees prepared in about two hours, and some others took six hours to prepare to leave. The area under the evacuation orders is smaller than area of a hurricane warning, which can vary by many square miles and involve millions of people and thousands of households (Figure 3.10). The logistical and transportation delays only grow larger when everyone

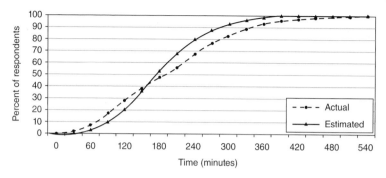

Figure 3.10 Cumulative distribution of evacuation preparation times. *Source:* Lindell *et al.,* 2007:54. Reproduced with permission from ASCE.

tries to evacuate at once. Lindell *et al.* (2007:54) find that officials recognize the significance of logistical preparation, but households also mention that they need time for psychologically preparing themselves to evacuate. Lindell and his colleagues also documented 10- to 20-hour traffic jams, and the time people say they need to get ready to go and to finally leave (Lindell *et al.*, 2007). This information can help decision makers decide when to put out warnings.

Updates from the National Hurricane Center in cone shapes showing time/space are provided every six hours and more often when conditions warrant (Lindell *et al.*, 2007:52). Figure 3.11 is an example of a probabilistic forecast in cone format from the National Hurricane Center for Hurricane Irene in 2011. This is one source of information that households use to make an evacuation decision. Lindell *et al.* point out that evacuation planners must recognize that people are not always home, not always together, and are not just sitting around waiting for the event or warning (2007:55).

Dr. Alan Stewart is a psychology professor at the University of Georgia who has always loved weather and is earning a degree in meteorology. In 2010 he was funded to teach schoolteachers in Georgia what they should teach their students about the science of weather and how to respond safely when it threatens in a weeklong series of workshops. The goal was to provide professional development to teachers about both the science and safety practices concerning hazards that are routinely faced in Georgia including lightning, tornadoes, hurricanes and floods. He used the Masters of Disaster

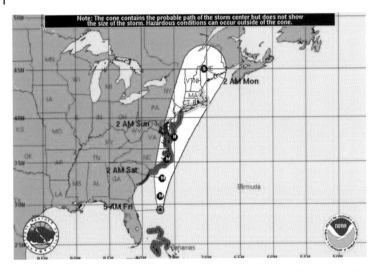

Figure 3.11 Example of a probabilistic forecast in cone format from the National Hurricane Center for Hurricane Irene in 2001. A graphic like this one is one of the sources of information that households use to make an evacuation decision. *Source:* National Hurricane Center, http://www.nws.noaa.gov/om/hurricane/ ww.shtml NOAA.

(K-8) curriculum that has been developed by the American Red Cross as a curricular resource.

Stewart also has conducted surveys that show that people living near the U.S. Gulf Coast may systemically underestimate the damages that some hurricanes can produce when landfall occurs. However, he found that if people are given additional information on the destructive potential of certain hurricanes, they would be more likely to evacuate compared to those who were given the hurricane category alone. "Hurricane destructiveness increases nonlinearly with increases in storm intensity. Our studies suggest that people are not aware of this relationship" (Stewart, 2011).

Other psychologists engaged in weather studies include Dr. Susan Joslyn, professor of psychology at University of Washington, and her colleagues. Their work focuses on ways to improve how uncertainty is communicated and increases understanding of how boaters and others make weather-sensitive decisions (Joslyn and Leclerc, 2012; Joslyn *et al.*, 2013; Joslyn *et al.*, 2009a, 2009b; Joslyn and Nichols, 2009;

Leclerc and Joslyn, 2012, 2015; Savelli and Joslyn, 2012) Dr. Rebecca Pliske, a cognitive psychologist, studied the decision-making processes of weather forecasters (Pliske *et al.*, 2004).

3.7 Sociology

Sociology is the "study of human behavior in society." Sociology is the study of the ways that individuals and groups interact within a society. Sociology studies the context and patterns in which human behavior occurs (Anderson and Taylor, 2007:2). Sociology is a scientific way of thinking about society and its influence on human groups (Anderson and Taylor, 2007:2). Sociology pays particular attention to established relationships between humans and institutions. Sociology is the study of the recurrent or regular aspects of human behavior (Jenkins, 2002:15). Some sociologists examine large social organizations such as businesses and governments, looking at their structures and hierarchies. Sociologists also look at divisions and inequality within society, examining phenomena such as race, gender, and class, and their effect on people's choices and opportunities.

In 2017, Dr. Lori Peek began her tenure as the Director of the Natural Hazards Center at the University of Colorado in Boulder. For many years, Dr. Peek was professor of sociology and co-director of the Center for Disaster and Risk Analysis at Colorado State University. She was the associate chair for the Social Science Research Council Task Force on Hurricane Katrina and Rebuilding the Gulf Coast. Peek's research focuses on marginalized populations and disasters. She studies how various forms of social inequality—such as those based on race, religion, gender, class, age, and ability—play out in people's everyday lives and during times of disaster. Her work uses predominantly qualitative methods, rather than quantitative. She also studies how unequal access to power and resources contributes to vulnerability before, during and after a disaster.

Following Hurricane Katrina in 2005, several hundred social science research studies were done. In 2009, Erikson and Peek assembled a 68-page research bibliography on the human effects of the storm that includes sociological as well as other impacts, including sections on children and schools, culture and tradition, displaced persons, economic effects and employment, elderly, emergency preparedness and response, environmental effects, evacuation, gender, health and health care, housing, media, post-disaster recovery, race and class, research methods, general books/edited volumes/overview articles, reviews, special issues of journals, documentary films, and select websites (http://wsnet.colostate.edu/CWIS584/Lori_Peek/hurricane-katrina-research-bibliography.aspx).

Many sociologists considered the status of the city New Orleans and its population before and after the storm. They studied how race, class, and capital affected vulnerable households before the storm and how these factors affected survivors after the storm. Results showed that people from a well-to-do background recovered quickly. Race and class are also shown to affect how a person viewed the aftermath, both in terms of the government's response and the broader social response. Historically disadvantaged groups saw prejudice. They studied how children were affected at the time of the storm and many years after (Peek and Fothergill 2011; Peek *et al.*, 2006).

Physical scientists, who study the characteristics of the hurricane, are often not aware of the challenges faced by the people who are vulnerable to the storms. Hurricane Katrina forced the largest and most abrupt displacement in U.S. history. About 1.5 million people evacuated from the Gulf Coast preceding Katrina's landfall. On Monday, August 29, 2005, Hurricane Katrina devastated more than 90,000 square miles of the U.S. Gulf Coast and, when the levee system gave way, drowned the city of New Orleans. The storm impacts destroyed or made unlivable approximately 300,000 homes and severely damaged 150,000 businesses. Nearly half of the estimated 110,000 people who remained in New Orleans did so because they did not believe the storm would be as bad as forecasted (Weber and Peek, 2012). The full force of Katrina did bypass the city, but the miles of weak levees could not withstand the storm surge.

The vast majority of those who either chose or were forced to stay behind were African American, poor, elderly, and/or living with a disability. As the levees collapsed and the city began to fill with water, children and adults sought refuge in attics, on rooftops, and on highway overpasses and other patches of dry ground. As many as 60,000 eventually made their way to the Superdome and the New Orleans Convention Center, often with the help of other stranded survivors. Damages are estimated at between $80 and $200 billion. The impacts killed 1720 initially and hundreds more in the months following from suicide, drug overdoses and other indirect causes.

Weber and Peek's edited collection of papers, *Displaced*, concerns the Katrina Diaspora. "Diaspora" refers to a scattering of the New

Orleans population around the country following the storm when the city was uninhabitable. Following the hurricane and the destruction of the City of New Orleans, the federal government relocated the residents all over the country. *Displaced* tells the stories of the lives of Katrina evacuees covering research in 13 communities in seven states across the United States. The book's chapter contributors describe the struggles that evacuees have faced in securing life-sustaining resources and rebuilding their lives. It also discusses the impacts that the displaced have had on communities that initially welcomed them and then later experienced "Katrina fatigue" as ongoing needs of evacuees strained local resources. *Displaced* reveals that Katrina took a particularly heavy toll on households headed by low-income, African American women who lost the support provided by local networks of family and friends. It also shows the resilience and resourcefulness of Katrina evacuees who have built new networks and partnered with community organizations and religious institutions to create new lives in the diaspora (Peek, 2012). Weber and Peek's book is written mostly by and primarily focuses on women.

"After awaiting relief and rescue, people were forced, sometimes at gunpoint, to evacuate. They were loaded into buses and airplanes and taken away, almost always with no idea of where they would end up. Although at first most of Katrina's evacuees sought and received shelter close to home, in the weeks following the hurricane, evacuees were scattered across all fifty states" (Weber and Peek, 2012).

The book reports research from 2005 to 2011 that included 767 in-depth interviews: 562 with displaced persons, 104 with first responders, service providers, and community organizers; and 101 with other residents in the receiving communities. Many of the studies employed multiple methods, including open and close-ended interviews, document analysis, participant observation, and focus groups. To capture the flow of respondents' unfolding lives, seven of the studies followed respondents over time (Weber and Peek, 2012). During Hurricane Katrina, thousands of people were rescued from rooftops. Peek states that this sort of trauma leads to a prolonged recovery period after a disaster. Peek's 2012 book *Displaced: Life in the Katrina Diaspora* includes an interview with an 18-year-old African American boy in New Orleans, who still broke down in tears four years after Katrina. "He was only 14 when Katrina struck. He lives with the memory of evacuating his grandmother in a wheelchair and carrying his baby sister in his arms, and then waiting for five days for somebody to rescue them.

Figure 3.12 Dr. Lori Peek conducting fieldwork in New Orleans after Hurricane Katrina. *Source:* Lori Peek. *Source:* A. Fothergill. Reproduced with permission from Dr. Peek.

He was displaced to three different cities in three different states after Katrina, and ultimately never returned to school (Figures 3.12 and 3.13). His experience of Katrina was profoundly different than someone who may have also lost their home, but who was able to self-evacuate, get out of harm's way, and return to a more stable routine soon after the disaster" (Peek, 2012). David and Enarson's edited 2012 collection, *The Women of Katrina How Gender, Race, and Class Matter in an American Disaster,* also provides an in-depth study of the intersection of weather and sociology.

Dr. Phaedra Daipha's 2015 book examines forecaster "screenwork" at the intersection of neuroscience and sociology (Daipha, 2015; Daipha, 2012; Daipha, 2010:150). *Masters of Uncertainty: Weather Forecasters and the Quest for Ground Truth* (2015) draws upon an ethnography of forecasting operations at the National Weather Service to develop a conceptual framework for studying uncertainty management in action.

Figure 3.13 Dr. Alice Fothergill and Dr. Lori Peek doing participatory research in New Orleans after Hurricane Katrina. *Source:* J. Tobin-Gurley. Reproduced with permission from Dr. Peek.

She describes how forecasters, as experts, use the information that they have on their screens in front of them and how forecasting developed as an indoor scientific activity far removed from observing the outside weather. Daipha discusses how forecasters present probabilistic statements about the future in quantitative ways as if the information is unambiguous or being presented to "imagined lay persons" or the "public." She acknowledges that forecasters have to cater to a wide variety of audiences with wildly different weather sensitivities (Daipha, 2015).

Daipha points out that there are "proactive" weather forecast users such as building contractors, farmers, and commercial fisherman who are consistently involved in weather-dependent activities where most other people have a more passive relationship to the weather and the forecasts. She concludes with a call for interdisciplinary research that "is not only event driven but fundamentally informed by a shared commitment to unearthing the forces inexorably drawing together uncertainty and expertise. Only then can we hope to proactively respond to the challenges of an uncertain future" (Daipha, 2012:22).

Sociologists Dr. Nicole Dash and Dr. Betty Morrow studied how the official delayed re-entry back home after a hurricane influences peoples' willingness to evacuate prior to the storm. While to many meteorologists, the decision to evacuate or not to evacuate when a hurricane warning goes into effect is a simple one. Sociologists' and psychologists' research show the complexity of the decision at the household and individual levels. Their research highlights perceived time constraints and can improve planning and forecast effectiveness. Social science case studies broaden understanding of what is involved in the publics' hurricane evacuation decisions that can help make warnings more realistic in terms of time necessary to evacuate and effectiveness (Lazo *et al.*, 2015).

The time frames for hurricane evacuation decision making are measured in days and in large spatial areas, especially when compared with the lead time and impact area of tornadoes. Within the days of lead time, each minute consists of a decision point for emergency managers, transportation planners, political officials and vulnerable populations (Dash and Gladwin, 2007). Dash and Morrow's (2001) research showed a surprising conclusion. In addition to the threat of the hurricane, potential evacuees consider the problem of delayed re-entry access to home or neighborhood after the evacuation and the storm. They also found that people fear that there will be long traffic delays, and this also affects their willingness to evacuate in the first place. After Hurricane George, evacuees were anxious to get home to evaluate damage and to prevent further damage. Dash and Morrow showed how blockades were set up by authorities, who thought the road was not safe for everyone to return after the storm, caused great frustration. In some cases, people waited 24 hours at the blockade before they were allowed to return home (2001:123). Drs. Dash and Morrow focussed on 1) the effects on future evacuation plans of experiencing return delays after Hurricane George and 2) the effects of the extensive media coverage of the delays on those who did not personally experience them (2001:120).

Dash and Morrow explain the "staged" evacuation system that calls for subpopulations to evacuate at different times based on what emergency managers' plan. For example, people in mobile homes and people with high medical needs are evacuated first (2001:121).

Sociologists have made a strong contribution in weather and disaster studies. Dr. Betty Morrow has worked closely with the NWS for decades helping them design warning products for storm surge that will effectively communicate risks and prompt vulnerable people to

take appropriate and timely actions (Morrow, 2009 and http://gss.
fiu.edu/people/faculty-emeriti/betty-morrow/). Dr. Alice Fothergill,
Professor of Sociology at University of Vermont, studied women in
the 1997 Grand Forks, ND flood (2004). Dr. Joseph Trainor and his
colleagues at University of Delaware have studied how false alarms
affect public willingness to take action (Nagele and Trainor, 2012).

Sociologist Dr. Gary Fine completed an ethnography of forecasters
in Chicago, Illinois in 2007 (Fine, 2007). In 2001 and 2002, he spent
hundreds of hours with various National Weather Services offices,
including the Norman Storm Prediction Center, the National Severe
Storms Laboratory, and three forecast offices. *Authors of the Storm*
picks up on the story telling side of social scientists. Sociologists, like
anthropologists, can be ethnographers and tell the stories of organizations or cultures. How do the meteorologists forecast the weather?
Based on his field research, Dr. Fine shows that the culture inside local
offices affects weather forecasts as much as any approaching weather.

In his book, Dr. Fine pays close attention to how weather "data" is
created as part of the process of producing knowledge in its social and
organizational context and how it is interpreted as well. There is a
considerable amount of debate and negotiation taking place both
within an office and between offices, in what Fine calls "the negotiated
order of collective knowledge" (p. 143). Fine points out the political
and social pressures the forecasters face, including having to issue
forecasts within very formal geographic constraints. For example, he
sees how forecasters deal with "the boxology" of tornado watches
along state boundaries. Forecasters use boxes to show where a tornado is expected, and they decide which box warrants a tornado watch
and which a tornado warning. Fine also discusses how forecasters deal
with the pressure to issue more detailed forecasts for urban centers
than for rural areas. Some people in urban areas do not appreciate
having their TV shows preempted for a weather warning for one of
the boxes for only outlying counties (Fine, 2007).

All six of these social sciences see weather and society through different, important lenses. They use similar scientific methods. They ask
related but particular questions, and they consider their data in different ways. The studies mentioned in this chapter are a small representation of the growing set of social science and weather research. When
early career social scientists and meteorologists engage in collaborations, there is a strong potential for meaningful progress in the challenges to reduce weather related loss of life and property damages.

3.8 Questions for Review and Discussion

1 Do you watch local television weather coverage? Do you check the weather forecast every day? If not, how often do you check the forecast? Are you aware of local mesonets that show real-time conditions near where you live? Do you check local mesonets or other observation stations to help you make decisions about what to wear or whether or not to ride your bicycle to work or school? If you do not check the weather forecasts, why not? Are you doubtful about the accuracy of the forecasts or are you a person who looks out the window and does not change your behavior based on a forecast, but reacts to whatever conditions present themselves as they occur? How do your friends and family members, who aren't as weather savvy as you, engage with formal and informal forecasts?

2 When you think about weather forecasts for your local area, what characteristics are most important to you? Do you care about the numbers such as high temperatures or wind speeds? Do you just want to know if it's a "nice" or a rainy day? Does your answer differ if you are on vacation or if you are working? Do you agree with the value-laden definitions of "bad" weather meaning days where being outside is unpleasant because of cold, heat, wind, or precipitation?

3 Do you think primarily of the physical characteristics such as how much it will rain or whether the water will flow out of the local waterway? Do you think about the impacts of the weather—that "bad weather" may affect the length of your commute to school or work? Does your answer to these questions vary depending on the day of the week or the particular responsibilities you have as a pet owner, landscaper, coach, or parent? Did your answer include both the physical elements of the weather and societal impacts? If you drive a car, do you ever drive through flooded roads? Do you think you tend to overestimate or underestimate severe weather risks to your personal safety? Are you personally prepared for severe weather? Do you have a weather radio with batteries, renter's insurance, and surge protectors for your electronics?

4 Do you think your demographic characteristics affect how much you change your behavior based on weather forecasts or weather conditions? These would include your gender, age, means of

transportation, length of time you've spent in your current environment, your type of work, and experience with severe weather. In what ways do you change your daily activities based on weather forecasts? Have you ever changed your route or means of transportation based on a forecast? Explain.

5 What suggestions do you have for improving weather forecasts? Do you prefer graphical presentations and maps or do you prefer text? Do you pay attention to longer-term forecasts beyond one to three days? Do your planned activities influence your interest in the forecast, say, if you are going on vacation or planning to do some gardening?

6 If you were in a mobile home and you heard sirens, what actions would you take? What factors would influence how you answer this question? Would you consider your transportation options and nearby shelters?

7 If humans could modify the weather in ways that shifted severe weather (hurricanes or tornadoes) or weather extremes (such as very icy conditions) away from heavily populated areas, what would be the pros and cons? How would "uncertainty" be considered?

8 If you were a cruise ship director and a hurricane forecast showed direct impact on your boat's route, how would you alter the cruise destinations? What steps would you need to take to notify the people who paid for the vacation cruise—when they signed the contract months before their trips? Would you have a caveat saying that since they were planning to travel during hurricane season that alternate routes might be necessary? Review the policies of two major cruise lines to support your response. How would you feel if the destinations and itinerary were altered if this was your first vacation cruise?

9 In 2005, more than 80% of the vulnerable population in New Orleans, Louisiana evacuated prior to the worst impacts of Hurricane Katrina. Was this a successful evacuation? What four factors would you take into account as part of your answer? What did you consider successful and what elements did you consider unsuccessful?

10 Special needs populations include people who are cognitively and/or physically disabled or frail to the point that they require routine, professional help to complete daily life tasks. This group can include patients and residents of hospitals, nursing homes, and mental health facilities that need additional specialized care during a disaster or evacuation activity. Much of this group is made up of the elderly who need assistance in daily living activities in addition to medical care. Do you agree that there should be a national registry of people with special needs (Fairfax County VA, 2016)? Who would be included in this registry? Does this definition work?

11 Insurance covers a great deal of damages for policyholders after many severe weather events. In the United States, flood insurance is separate from homeowners' insurance. As weather extremes continue to be exacerbated by climate change, will insurance companies be able to continue to pay out claims? There already have been events where insurance companies have paid out for a hurricane like Hurricane Andrew in 1992 and then would not insure Florida residents who tried to renew their policies. This also was the case following major wildfires where residents were denied new policies. What are the ethics of these actions? How would your opinion vary if you were a homeowner or if you were a shareholder in an insurance company?

3.9 Using What You've Learned: Homework Assignment From the Chapter

1 The references and appendices of this textbook contain numerous scholarly articles on the topic of weather and society. Familiarize yourself with the work that already exists in the developing field of integrated socio-meteorological research by searching the lists of references and to find some of the articles and books that appeal to your interests. Once you have read a few articles, you will be able to take a critical eye to the research that is available. What topics do the references leave out that should be included? Are the case studies relevant?

2 Select an article from the references of this text or one published in a meteorology or hazards journal in the last two years that you find on your own. Critically examine the research. Who are the authors? What are their disciplinary backgrounds? What research questions are they asking? How do their disciplinary backgrounds inform the types of questions they are asking? Is their disciplinary background the best equipped to answer those questions? What methodologies did they use? What are the limitations of those methodologies? How large was their sample size? Is their research generalizable, or is it more of an initial step that can be built upon with future research? Do the authors recognize the limitations of their research? How do the authors present their findings? For example, do they quantify their results in tables and graphs, generate theoretical models based on their findings, or present a narrative? Is this presentation the best way of sharing the data? What types of conclusions do they come to and are the conclusions supported by the data presented in the article? Write a one-page essay discussing the article and your answers to the questions.

References

Anderson, M.L., and Taylor, H.F. (2007) *Sociology: Understanding a Diverse Society*. New York: Thomson Wadsworth.

Andrey, J. (2010) Long-term trends in weather-related crash risks. *Journal of Transport Geography*, 18: 247–258.

Bernstein, D., Penner, L.A., Clarke-Stewart, A., and Roy, E. (2011) *Psychology*. New York: Cengage Advantage Books.

Brinkman, B. (2013) bobbrinkmann.blogspot.com/2013/09/a-tale-of-two-hurricaneguides.html?utm_source=feedburner&utm_medium=email&utm_campaign=Feed%3A+blogspot%2FCDVYa+%28On+the+Brink%29&utm_content=Yahoo!+MailIn (accessed July 23, 2017).

Craft, E.D. (2010) An economic history of weather forecasting. http://eh.net/encyclopedia/article/craft.weather.forcasting.history (accessed July 23, 2017).

Cutter, S.L., Johnson, L.A., Finch, C., and Berry, M. (2007) The U.S hurricane coasts: Increasingly vulnerable? *Environment*, 49: 8–21.

Cutter, S. (2013) Preparing for the worst should be a nonpartisan issue. http://www.thestate.com/2013/08/21/2931514/cutter-preparing-for-the-worst.html#storylink=cpy (accessed July 23, 2017).

Cutter, S.L., and Smith, M.M. (2009) Fleeing from the hurricane's wrath: Evacuation and the two Americas. *Environment*, 51: 26–36.

David, E., and Enarson, E. (2012) *The Women of Katrina How Gender, Race, and Class Matter in an American Disaster*. Nashville, TN: Vanderbilt University Press.

Daipha, P. (2015) *Masters of Uncertainty: Weather Forecasters and the Quest for Ground Truth*. Chicago: University of Chicago.

Daipha, P. (2012) Weathering risk: Uncertainty, weather forecasting, and expertise. *Sociology Compass*, 6(1): 15–25.

Daipha, P. (2010) Visual perception at work: Lessons from the world of meteorology. *Poetics*, 38(2): 150–164.

Dash, N., and Gladwin, H. (2007) Evacuation decision making and behavioral responses: Individual and household *Natural Hazards Review*, 8: 69–77.

Dash, N., and Morrow, B.H. (2001) Return delays and evacuation order compliance: The case of Hurricane Georges and the Florida Keys. *Environmental Hazards*, 2: 119–128.

Demuth, J.L., Morss, R.E., Morrow, B.H., and Lazo, J.K. (2012) Creation and communication of hurricane risk information *Bulletin of the American Meteorological Society*, 93: 1133–1145.

Demuth, J.L., Morss, R.E., Lazo, J.L., and Hildebrand, D.C. (2013) Improving effectiveness of weather risk communication on the NWS point-and-click web page. *Weather and Forecasting*, 28: 711–726.

Dixon, G., Clarke, C., Eosco, G., Holton, A., and Mckeever, B. (2015) The power of a picture: Overcoming scientific misinformation by communicating weight-of-evidence information with visual exemplars *Journal of Communication*, 65(4): 639.

Ellis, L. (1994) *Research Methods in the Social Sciences*. New York: Brown and Benchmark.

Eosco, G. (2015) Town Hall Meeting: Watch out! A review of the National Weather Service's watch, warning, advisory hazard messaging system. It's advised you attend. You have been warned! American Meteorological Society meeting, Phoenix, AZ, January 5.

Fairfax County, VA Fairfax alerts – Functional needs. (2016) http://www.fairfaxcounty.gov/specialneeds/ (accessed July 23, 2017).

Fine, G.A. (2007) *Authors of the Storm Meteorologists and the Culture of Prediction.* Chicago: University of Chicago.

Ford, J.D., Pearce, T., Duerden, F., Furgal, C., and Smit, B. (2010) Climate change policy responses for Canada's Inuit population: The importance of and opportunities for adaptation. *Global Environmental Change*, 20(1): 177–119.

Fothergill, A. (2004) *Heads Above Water: Gender, Class, and Family in the Grand Forks Flood.* New York: State University of New York Press.

Freberg, K., Saling, K., Vidoloff, K.G., and Eosco, G. (2013) Using value modeling to evaluate social media messages: The Case of Hurricane Irene. *Public Relations Review*, 39(3): 185–192.

Frei, T. (2010) Economic and social benefits of meteorology and climatology in Switzerland. *Meteorological Applications*, 17: 39–44.

Friedman, M., and Savage, L.J. (1948) The utility analysis of choices involving risk. *Journal of Political Economy*, 56: 279–304.

Gigerenzer G., Hertwig, R., van den Broek, E., Fasol, B., and Katsikopoulos, K.V. (2005) A 30% rain tomorrow: How does the public understand probabilistic weather forecasts? *Risk Analysis*, 25: 623–629.

Gladwin H., Lazo, J.K., Morrow, B.H., Peacock, W.G., and Willoughby, H.E. (2007) Social science research needs for the hurricane forecast and warning system. *Natural Hazards Review*, 8: 87–95.

Gladwin, C., Gladwin, H., and Peacock, W. (2001) Modeling hurricane evacuation decisions with ethnographic methods. *International Journal of Mass Emergencies and Disasters*, 19: 117–143.

Gladwin, H., and Peacock, W. (1997) Warning and evacuation: A night for hard houses. In: Peacock *et al.* (eds.), *Hurricane Andrew: Ethnicity, Gender, and the Sociology of Disasters.* London: Routledge, pp. 52–74.

Hayden, M., Drobot, S., Gruntfest, E., Benight, C., Radil, S., and Barnes, L. (2007) Information sources for flash flood warnings in Denver, CO and Austin, TX. *Environmental Hazards*, 7: 211–219.

Jenkins, R. (2002) *Foundations of Sociology Towards a Better Understanding of the Human World.* New York: Palgrave Macmillan.

Joslyn, S.L., and LeClerc, J.E. (2012) Uncertainty forecasts improve weather-related decisions and attenuate the effects of forecast error. *Journal of Experimental Psychology: Applied*, 18(1): 126.

Joslyn, S., Nemec, L., and Savelli, S. (2013) The benefits and challenges of predictive interval forecasts and verification graphics for end users. *Weather, Climate, and Society*, 5(2): 133–147.

Joslyn S., Nadav-Greenberg, L., and Nichols, R.M. (2009a) Probability of precipitation: Assessment and enhancement of end-user understanding. *Bulletin of the American Meteorological Society*, 90: 185–193.

Joslyn S., Nadav-Greenberg, L., Taing, M.U., and Nichols, R.M. (2009b) The effects of wording on the understanding and use of uncertainty information in a threshold forecasting decision *Applied Cognitive Psychology*, 23: 55–72.

Joslyn, S.L., and Nichols, R.M. (2009) Probability or frequency? Expressing forecast uncertainty in public weather forecasts *Meteorological Applications*, 16: 309–314.

Kates, R.W., Colten, C.E., Laska, S., and Leatherman, S.P. (2006) Reconstruction of New Orleans after Hurricane Katrina: A research perspective. *Proceedings of the National Academy of Sciences*, 103(40): 14653–14660.

Katz, R., and Murphy, A. (1997) *Economic Value of Weather and Climate Forecasts*. Cambridge, UK: Cambridge University Press.

King, S.S. (1989) *Human Communication as a Field of Study: Selected Contemporary Views*. New York: State University of New York Press.

Kox, T., Gerhold, L., and Ulbrich, U. (2015) Perception and use of uncertainty in severe weather warnings by emergency services in Germany. *Atmospheric Research*, 158-159(1-15): 292–301.

Laska, S., and Morrow, B. (2006) Social vulnerabilities and Hurricane Katrina: An unnatural disaster in New Orleans. *Journal of the Marine Technology Society*, 40(4): 16–26.

Lazo, J.K. (2010) The costs and losses of integrating social sciences and meteorology. *Weather, Climate and Society*, 2: 171–173.

Lazo, J.K., Morss, R.E., and Demuth, J.L. (2009) 300 billion served: Sources, perceptions, uses, and values of weather forecasts. *Bulletin of the American Meteorological Society*, 90: 785–798.

Lazo, J., Bostrom, A., Morss, R.E., Demuth, J., and Lazrus, H. (2014) *Communicating hurricane warnings: Factors affecting protective behavior*. Conference on Risk, Perceptions, and Response. Boston, MA: Harvard University.

Lazo, J.K., Lawson, M., Larsen, P., and Waldman, D. (2011) Sensitivity of the U.S economy to weather variability. *Bulletin of the American Meteorological Society*, 92: 709–720.

Lazo, J.K., Bostrom, A., and Morss, R.E. *et al.* (2015) Factors affecting hurricane evacuation intentions. *Risk Analysis*, 35(10): 1837–1857.

LeClerc, J., and Joslyn, S. (2012) Odds ratio forecasts increase precautionary action for extreme weather events. *Weather, Climate, and Society*, 4(4): 263–270.

LeClerc, J. and Joslyn, S. (2015) The cry wolf effect and weather-related decision making. *Risk Analysis*, 35(3): 385–395.

Letson, D., Sutter, D.S., and Lazo, J.K. (2007) Economic value of hurricane forecasts: An overview and research needs. *Natural Hazards Review*, 8: 78–86.

Lindell, M.K., Lu, J., and Prater, C.S. (2005) Household decision-making and evacuation in response to Hurricane Lili. *Natural Hazards Review*, 6: 171–179.

Lindell, M.K., Prater, C.S., and Peacock, W. (2007) Organizational communication and decision making for hurricane emergencies. *Natural Hazards Review*, 8: 50–60:

McNeeley, S., and Lazrus, H. (2014) The cultural theory of risk for climate change adaptation. *Weather, Climate, and Society*, 6: 506–519.

Morrow, B.H. (2009) *Improving Coastal Risk Communication: Guidance from the Literature Report*. Prepared for NOAA's Coastal Service Center, Charleston, SC. http://www.csc.noaa.gov/Risk_Behavior_&_Communication_Report.pdf (accessed July 23, 2017).

Morss, R.E., Demuth, J., and Lazo, J. (2008) Communicating uncertainty in weather forecasts: A survey of the U.S public. *Weather and Forecasting*, 23: 974–991.

Morss, R.E., Wilhelmi, O.V., Downton, M., and Gruntfest, E. (2005) Flood risk, uncertainty, and scientific information for decision-making: Lessons from an interdisciplinary project. *Bulletin of the American Meteorological Society*, 86: 1593–1601.

Murray, K.B., Di Muro, F., Finn, A., and Popkowski Leszczyc, P. (2010) The effect of weather on consumer spending. *Journal of Retailing and Consumer Services*, 17(6): 512–520.

Nagele, D.E., and Trainor, J.E. (2012) Geographic specificity, tornadoes, and protective action. *Weather, Climate, and Society*, 4(2): 145–155.

NOAA Science Advisory Board Social Science Working Group. (2009) *Integrating Social Science into NOAA Planning, Evaluation and Decision making a Review of Implementation to date and Recommendations for Improving Effectiveness*. A Report from the NOAA Science Advisory Board, April 16.

NOAA. (2011) National Weather Service. http://www.nws.noaa.gov/com/files/weatherreadynation101presentation.pdf (accessed July 23, 2017).

Oliver-Smith, A. (1996) Anthropological research on hazards and disasters. *Annual Review of Anthropology*, 25: 303–328.

Oliver-Smith, A., and Hoffman, S.M. (1999) *The Angry Earth: Disaster in Anthropological Perspective*. London: Routledge.

Peek, L. (2012) *Displaced: Life in the Katrina Diaspora*. Austin: University of Texas Press.

Peek, L., and Fothergill, A. (2011) Using focus groups: Lessons from studying daycare centers, 9/11, and Hurricane Katrina *Qualitative Research*, 9(1): 31–59.

Peek, L., Morrissey, B., and Marlatt, H. (2006) Disaster hits home: A model of displaced family adjustment after Hurricane Katrina. *Journal of Family Issues*, 32(10): 1371–1396.

Pennesi, K., Arokium, J., and McBean, G. (2012) Integrating local and scientific weather knowledge as a strategy for adaptation to climate change in the Arctic. *Mitigation and Adaptation Strategies for Global Change*, 17(8): 892–922.

Pennesi, K. (2011) Making forecasts meaningful: Explanations of problematic predictions in northeast Brazil. *Weather, Climate and Society*, 3: 90–105.

Pennesi, K. (2007) Improving forecast communication: Linguistic and cultural considerations *Bulletin of the American Meteorological Society*, 88: 1033–1044.

Peterson, N., and Broad, K. (2008) Climate and weather discourse in anthropology: From determinism to uncertain futures. In: Crate, S., and Nuttall, M. (eds.), *Anthropology and Climate Change: From Encounters to Actions*. Portland: Left Coast Press, pp. 70–86.

Pliske, R., Crandall, B., and Klein, G. (2004) Competence in weather forecasting. In: Shanteau, J., and Johnson, P.E., (eds.), *Psychological Investigations of Competence in Decision Making*. Cambridge, UK: Cambridge University Press, pp. 40–68.

Pollard, J.S., Oldfield, J., Randalls, S., and Thornes, J.E. (2008) Firms, weather derivatives and geography. *Geoforum*, 39: 616–624.

Ruin, I., Creutin, J.-D., Anquetin, S. and Lutoff, C. (2008) Human exposure to flash floods – Relation between flood parameters and human vulnerability during a storm of September 2002 in Southern France. *Journal of Hydrology*, 361: 199–213.

Ruin, I., Gaillard, J.-C., and Lutoff, C. (2007) How to get there? Assessing motorists' flash flood risk perception on daily itineraries Advances and challenges in flash flood warnings *Environmental Hazards*, 7: 235–244.

Ruin I., Lutoff, C., Boudevillain, B., Creutin, J.-D., Anquetin, S., Bertran Rojo, M., Boissier, L., Bonnifait, A., Borga, M., Colbeau-Justin, L., Creton-Cazanave, L., Delrieu, G., Douvinet, J., Gaume, E., Gruntfest, E., Naulin, J.-P., Payrastre, O., and Vannier, O. (2014) Social and hydrological responses to extreme precipitations: An interdisciplinary strategy for post-flood investigation. *Weather, Climate, and Society*, 6: 1135–1153.

Savelli, S., and Joslyn, S. (2012) Boater safety: Communicating weather forecast information to high-stakes end-users. *Weather, Climate and Society*, 4: 7–19.

Schumacher, R.S., Lindsey, D.T., Schumacher, A.B., Braun, J., Miller, S.D., and Demuth, J.L. (2010) Multidisciplinary analysis of an unusual tornado: Meteorology, climatology, and the communication and interpretation of warnings *Weather and Forecasting*, 25: 1412–1429.

Simmons, K.M., and Sutter, D. (2011) *Economic and Societal Impacts of Tornadoes*. Washington, DC: American Meteorological Society.

Simmons, K.M., and Sutter, D. (2009) False alarms, tornado warnings, and tornado casualties. *Weather, Climate, and Society*, 1: 38–53.

Smith, Alan H., and John L. Fischer, eds. (1970) *Anthropology*. Englewood Cliffs: Prentice-Hall.

Souza, K. (2012) Wal-Mart, big boxes brace for Sandy. October 29. http://www.thecitywire.com/node/24804#.UW2zlb-aGxF (accessed July 23, 2017).

Spinney, J., and Gruntfest, E. (2012) *What makes our partners tick? Using ethnography to inform the Global System Division's development of the Integrated Hazards Information Services (IHIS)*. Report prepared for NOAA Integrated Hazards Information Systems Project. http://www.evegruntfest.com/SSWIM/pdfs/Final-rep-1.pdf (accessed July 23, 2017).

Spinney, J.A., and Pennesi, K.E. (2012) When the river started underneath the land: Social constructions of a 'severe' weather event in Pangnirtung, Nunavut. *Canada Polar Record*, 1-11.

Stewart, A.E. (2011) Gulf coast residents underestimate hurricane destructive potential. *Weather, Climate and Society*, 3: 116–122.

Strauss, S. and Orlove, B. (2003) *Weather, Climate, Culture*. New York: Berg.

Sutter, D., and Erickson, S. (2010) The time cost of tornado warnings and the savings with storm-based warnings. *Weather, Climate, and Society*, 2: 103–112

Terti, G. (2014) Forecasting of flash flood impacts integrating the space-time distribution of social vulnerability. Powerpoint presentation. San Antonio, Texas, November 21.

Terti, G., Ruin, I., Anquetin, S., and Gourley, J. (2015) Dynamic vulnerability factors for impact-based flash flood prediction. *Natural Hazards*, 79(3): 1481–1497.

Trumbo, C., Lueck, M., Marlatt, H., and Peek, L. (2012) The effect of proximity to hurricanes Katrina and Rita on subsequent hurricane outlook and optimistic bias. *Risk Analysis*, 31: 1907–1918.

University Corporation for Atmospheric Research (UCAR). (2011) Press release. http://www2.ucar.edu/news/4810/economic-cost-weather-may-total-485-billion-us (accessed July 23, 2017).

Von Gruenigen, S., Wilemse, S., and Frei, T. (2014) Economic value of meteorological services to Switzerland's airlines: The case of TAF at Zurich airport. *Weather Climate and Society*, 6: 264–272.

Weatherhead, E.C., Gearheard, S., and Barry, R. (2010) Changes in weather persistence: Insight from Inuit knowledge. *Global Environmental Change*, 20: 523–528.

Weber, L., and Peek, L. (eds). (2012) *Displaced: Life in the Katrina Diaspora*. Austin: University of Texas Press.

4

Thirteen Profiles of Leaders in Weather and Social Science

This section introduces some of the pioneers who are passionately dedicating their careers to research and practice at the intersection of weather and society. These individuals represent different academic backgrounds, geographic locations, and areas of interest. Their backgrounds and their career trajectories are listed. Each person was asked to make recommendations for how he or she thought the social sciences and atmospheric science could achieve integration and their ideas are included. This section aims to address the inquiries from eager, bright, and ambitious students and others looking to build their career on work that integrates meteorology and social science. These profiles highlight the fact that there is no "correct" track to take to be a part of the movement to integrate social science into meteorology. Notice similarities and differences in their backgrounds. Each person is willing to answer questions about their professional journeys and how their experiences might inspire anyone interested in pursuing a similar career path.

Exciting examples of societally relevant meteorology are happening in the NWS field offices. Included in this list are two of many possible examples of NWS forecasters who are leading the way.

Angle, Kelsey   kelsey.angle@noaa.gov
Angle is the Warning Coordination Meteorologist at NOAA/NWS's Weather Forecast Office, Des Moines, Iowa (Figure 4.1).
Academic background: B.S. in Meteorology from University of Oklahoma, Certificate in Certified Public Management from University of Kansas

Angle began his career in the NWS in the Pacific Region at the Honolulu Weather Forecast Office and the Central Pacific Hurricane Center.

Weather and Society: Toward Integrated Approaches, First Edition. Eve Gruntfest.
© 2018 John Wiley & Sons Ltd. Published 2018 by John Wiley & Sons Ltd.

Figure 4.1 Kelsey Angle. *Source:* K. Angle.

He has worked in the Central Region as a Forecaster at Springfield, Missouri and a Senior Forecaster at the Topeka, Kansas Weather Forecast Offices. Prior to working for the NWS, Angle worked as a broadcast meteorologist in Wichita and Topeka, Kansas. Angle served as an Emergency Response Specialist from 2011-2015 at the NWS Central Region Operations Center, a component of the National Weather Service Weather-Ready Nation Strategic Plan. He coordinated with NWS Forecast Offices and provided impact-based decision support services (IDSS) to state, regional, and national partners ranging from state emergency management agencies to federal agencies such as the Federal Emergency Management Agency (FEMA), the United States Geological Survey (USGS), and the Environmental Protection Agency (EPA). During regional and national high-impact weather events, Angle served as a point of contact between the regional office and stakeholders by providing on-demand briefings, scheduled briefings, and multi-media reports incorporating weather and societal impact information to aid decision makers. Angle provided on-site support for Red River of the North Flooding, Missouri River Flooding, Mississippi River Flooding, and supported of numerous national special security events.

As program manager, Angle promoted IDSS regionally and provided programmatic support and guidance to Central Region field offices. He held monthly training webinar calls to further advance and enhance IDSS skills, tools, and services. Through NOAA's Rotational Assignment Program, Angle assisted in the development and facilitation of six NWS Deployment Boot Camps. Over 90 NWS meteorologists and hydrologists have participated in the Decision Support Services

Deployment Boot Camp at the NWS Training Center. This course is designed to build capacity and capability for providing quality on-site decision support services in critical situations. By offering a balanced mix of presentations, large- and small-group discussions, and facilitated interactive exercises, the group is challenged to clarify and practice skill sets needed to deliver quality DSS for high impact events.

Through interactive exercises and simulations with other agencies including FEMA, the United States Army Corp of Engineers (USACE), the EPA, and Emergency Managers, class members sharpen and strengthen their skill sets. Activities include providing Incident Command System (ICS)-style briefings, developing multimedia presentations, conducting media interviews, filing documentation, and utilizing social media and mobile technology tools in the context of incident support.

The NWS Central Region Operations Center also coordinates with local, state, and federal emergency management to plan and provide IDDS for high-profile, large-venue events of regional and national significance such as Major League All-Star Games, the NATO Summit in Chicago, the Super Bowl, and the Indianapolis 500. These events are often staffed with Emergency Response Specialists in coordination and collaboration with emergency management partners.

Working with FEMA, Angle co-developed and facilitated three Collaborative Weather Impact Assessment and Risk Communication Training Classes. Dozens of FEMA employees attended the class designed to enhance FEMA Watchstanders knowledge of weather concepts, NWS products and services, and resource navigation to improve weather information analysis and interpretation capabilities for impact assessment and dissemination during steady-state and monitoring operations through disaster activation.

In March 2009, Angle planned and organized the Southwest Missouri Flash Flood Integrated Warning Team Symposium. This inaugural Integrated Warning Team symposium was in coordination with the Southwest Missouri Council of Governments. Over 70 participants representing federal, state and local agencies, private businesses, members of the public, education and research institutions, and the media participated to foster open communication, joint partnerships through sharing of information and resources and to develop and incorporate mitigation and safety preparedness strategies regarding flooding in southwest Missouri and southeast Kansas. Angle has been involved in two other Integrated Warning Team workshops in Detroit, MI and Wichita, KS.

4.1 Angle's Recommendations for Integrating Social Science and Atmospheric Science

Angle highlights the importance of listening, observing, and seeking understanding of partner needs, uses of weather information, and their decision making. He comments "When there is a true understanding of the partner's reality, decision making, variables involved and timelines, impact-based weather, water and climate information communicated in a clear, concise, timely and user-friendly fashion can be worth gold to partners. Accurate, up-to-date, relevant, understandable, and actionable weather, water and climate information builds trust and results in successful, valuable, and serving partnerships."

Becker, Julia   J.Becker@gns.cri.nz
Dr. Becker is a research scientist at GNS Science Wellington, New Zealand
Academic Background: B.Soc.Sci. Geography, Earth Science and M.Sc. Earth Sciences from Waikato University, and Ph.D. in Psychology from Massey University (http://www.gns.cri.nz/who/staff/1986.html) (Figure 4.2).

Figure 4.2 Julia Becker. *Source:* Reproduced with permission from J. Becker.

In Dr. Becker's first undergraduate year at university, she picked up a number of subjects of interest including geography and earth science. She read every book and all the articles she could find about natural hazards, resources and environment, volcanology, and sedimentology. She particularly loved studying the human elements of risk and hazards. Her master's degree also followed this direction and involved researching how volcanic risk to people can be managed in the Waikato Region of New Zealand.

Dr. Becker then worked in a children's science center. This job helped her realize how important and difficult it is to communicate science to lay audiences. Her next employment was with GNS (Geological and Nuclear Sciences) as a researcher on natural hazards and society. Following working two years at GNS Science, she traveled overseas and worked as an environmental planning researcher at Oxford Brookes University in the U.K. for two years.

In 2012, Dr. Becker returned to New Zealand to another research position at GNS Science, where she currently works. GNS Science is part of the Joint Centre for Disaster Research in collaboration with Massey University. She continues to do research on how society is affected by and responds to natural hazards, including those related to weather. Her key areas of expertise include preparedness for disasters, community resilience, land-use planning for natural hazards, human response to warnings, and disaster response and recovery. She has worked with the New Zealand MetService on a project to help understand human response to short-fuse weather warnings and with the Australian New South Wales State Emergency Service on a number of flood perception and behavior projects. Becker completed a Ph.D. in psychology and was driven by her desire to understand human behavior in relation to natural hazards and to provide advice on how appropriate behaviors can be encouraged to reduce losses.

4.2 Dr. Becker's Recommendations for Integrating Social Science and Atmospheric Science

The integration of physical and social science remains a worthy challenge. Dr. Becker's view is that science should be able to help people. At GNS Science and Massey, she says:

"We have achieved success in integrating physical and social science by developing genuine research partnerships with others, including partnerships with other physical and social researchers, and partnerships with key agencies. In a research context, physical and social scientists should develop research projects together that can provide end-to-end coverage of issues and incorporate both scientific and human aspects. Close research partnerships with agencies (e.g., meteorological or emergency management agencies) are also essential to ensure that physical and social aspects are covered in a project, that real life issues are investigated, and that positive outcomes can be reached for all involved (including scientists, agencies and the public)."

Figure 4.3 Julie Demuth. *Source:* J. Demuth.

Demuth, Julie   jdemuth@ucar.edu

Dr. Demuth is a Research Scientist in the Mesoscale and Microscale Meteorology Lab at the National Center for Atmospheric Research (NCAR) in Boulder, Colorado (https://staff.ucar.edu/users/jdemuth) (Figure 4.3).

Academic Background: B.S. in Meteorology from University of Nebraska-Lincoln, M.S. in Atmospheric Science, and Ph.D. in Public Communication and Technology from Colorado State University

Dr. Demuth was born and raised in Nebraska. Her family was living in Grand Island on June 3, 1980, when seven tornadoes struck the town. She was too young to remember the tornadoes, but she grew up reading and re-reading all the newspapers and magazines that commemorated that event. Between that memory and her dad always taking her outside to watch the weather, it is easy to see why she, as she admits "is among those weather geeks who grew up knowing she wanted to be a meteorologist since she was old enough to say the word."

Dr. Demuth's first weather-related jobs were a series of internships in the Washington DC-area in the summers during her undergraduate years. She first interned with the NOAA Test and Evaluation Branch (that was located on the same grounds as the NWS Washington DC/ Baltimore Weather Forecast Office), where they quality-checked everything from radiosondes to the array of ASOS (Automated Surface Observing System) instruments. The following summers, she interned with TRW, a private contractor that was developing weather decision support systems such as the Weather and Radar Processor (WARP) and the Integrated Terminal Weather System (ITWS) for the airline sector.

During her senior year of her undergraduate program in meteorology, a professor suggested that she go to graduate school. Although she had never considered grad school before that moment, she started surfing the web (which was an adventure in and of itself, because in fall 1998, her internet was dial up and university webpages were quite rudimentary). After that search, she started applying to programs, taking the necessary pre-graduate school exams and visiting schools. She landed at Colorado State University (CSU) in the Department of Atmospheric Science. There, she was fortunate to work closely with Dr. Mark DeMaria for her master's thesis research using satellite data to study hurricanes. She enjoyed the experience and is immensely thankful for DeMaria's tutelage. Yet, it was during her master's program that she realized she wanted to do something without knowing what that something was, something more connected to the "human side" of weather.

After she finished her master's at CSU, she applied for and received the Christine Mirzayan Science and Technology Policy Fellowship with The National Academies (more commonly referred to as the National Academy of Sciences). She interned jointly with the Disasters Roundtable and the Board on Atmospheric Sciences and Climate (BASC). Upon finishing her four-month internship, she realized she wasn't ready to commit to pursuing a Ph.D. in atmospheric science. She ultimately was offered a permanent position with BASC, and she worked as a Program Officer there for two years doing science policy work.

While working at BASC, Demuth was greatly influenced by the passionate social scientists she met, including Dr. Eve Gruntfest. Then in 2004, Demuth and Gruntfest began a dynamic, 18-month, full-time partnership at NCAR that was crucial to the development of the WAS*IS workshops and movement, as well as Demuth's formative exposure to societal aspects of weather.

In the fall of 2009, she took the plunge and began a Ph.D. program in Colorado State University's Department of Journalism and Technical Communication. She completed the Ph.D. in 2015. In 2017, Demuth still works at NCAR as an Associate Scientist with the NCAR Mesoscale and Microscale Meteorology lab where she does research on perceptions and communication of weather risk information.

Dr. Demuth notes, "Although work and school keep me busy, I found my passion – after several steps and thanks to the wonderful people I met along the way – and I wouldn't have it any other way!"

4.3   Dr. Demuth's Recommendations for Integrating Social Science and Atmospheric Science

Dr. Demuth says, "The primary ingredient needed for integrating atmospheric and social sciences is the passion to do so. The pressing questions and challenges that exist are situated where the boundaries of atmospheric and social science disciplines intersect and overlap. This is where new, transformative ideas happen ... but it's also where things get messy and hard. You have to be willing to be frustrated and to think differently. You have to listen as much (or more!) than you speak, and respect the different sets of knowledge, experiences, and ideas of the different atmospheric and social science researchers, broadcast meteorologists, emergency managers, and many other practitioners you'll talk to and work with. You have to realize and accept that the way you've thought about and approached the world in the past might now have to change, perhaps slightly, perhaps significantly.

Finally, you have to realize this is still an emerging area of work, which poses its own challenges. I quickly realized when I started my Ph.D. program that everyone else in my department and at the social science conferences I attend (like the Society for Risk Analysis annual conference) isn't sitting around eager to talk about the weather all the time, including how different concepts and theories might be applied to or are different for weather risks. Instead, the topics we discuss are more along the lines of the real-world impacts of media violence, social media use during presidential campaigns, or people's perceptions about vaccine risks; then I have to think about if and how these might apply to weather. Thus, the vast majority of my thinking about how to integrate the mass communication and risk communication literature into the weather domain is done on my own. I'm fortunate that I have good work colleagues I can talk with, but most of it is still done on my own. All these challenges of integrating atmospheric and social sciences can be overcome if you have one thing—the passion to do this work."

Figure 4.4 Greg Dobson. *Source:*
Reproduced with permission from
G. Dobson.

Dobson, Greg   gdobson@unca.edu
Dobson is a Research Scientist and GIS Coordinator at University of
  North Carolina at Asheville's National Environmental Modeling
  and Analysis Center (NEMAC) (https://nemac.unca.edu/faces/
  staff/j-greg-dobson) (Figure 4.4).
Academic Background: B.S. in Geography from East Tennessee State
  University, M.A. in Geography from Appalachian State University

Dobson views himself as being "in the middle" of this intersection of
atmospheric and social science. Technically, he is neither a meteorologist nor a social scientist. His background is in geography.
He considers geography as "the wonderful field of study that lets
you study anything and everything." It was his love and passion for
all things weather and climate (having observed and questioned
many major weather events during his childhood), combined with his
interests in understanding how society responded to these events,
plus his lifelong passion for maps and understanding spatial processes
that led him to study geography, especially geographic information
systems (GIS) and to pursue a career in this field.

Dobson has served as NEMAC's GIS Coordinator for over 10 years.
He views GIS and geospatial technologies as the "common language"
that both physical and social scientists can speak and understand.
With geography at the core of his background and education, he
enjoys applying GIS processes to a wide range of topics, including
flood hazards, emergency response, weather and climate, climate

change, and enhanced visualization techniques. By taking basic data about precipitation, land use, topography, and other relevant elements, Dobson develops multi-layered maps that help decision makers to mitigate hazards. He believes that these decision makers, research scientists, and other stakeholders, including the general public, need to clearly understand the data in front of them before they will ever be able to fully appreciate the issues they need to address.

In 2009, Dobson attended the summer WAS*IS Workshop, and he considers that to have been a career-changing experience for him. However, having studied geography, he certainly had not fully understood the complex questions that arise when considering how society responds to weather/climate/hydro events and issues. After WAS*IS, Dobson more fully recognized the need to work from the bottom up with key stakeholders in an effort to better understand their needs and sensitivities. Dobson has been able to take this commitment to NEMAC and apply it to many of his current projects, including a large effort with NOAA and the U.S. Global Change Research Program in formulating the next National Climate Assessment. He works with a variety of stakeholders representing many different sectors, including energy, transportation, public health, environment, society, among many others, who were all interested in better understanding potential impacts resulting from increased climate variability and climate change. Dobson's experience is an excellent example of how WAS*IS concepts can be applied to these other closely related fields, such as climate and spatial understanding of societal impacts using GIS (Dobson and Boenhert, 2015).

4.4 Dobson's Recommendations for Integrating Social Science and Atmospheric Science

Atmospheric and social scientists must continue to become more aware of the others work and learn how to better integrate with each other. The bottom line is the need for better education, awareness, training opportunities, collaborative research, and professional interaction. Federal and academic funding will help, but a dedication to the "grass-roots" movement is also necessary.

Fransen, Tanja   tanja.fransen@noaa.gov
Fransen is the Meteorologist-in-Charge at NOAA/NWS's Weather Forecast Office, Glasgow, Montana (Figure 4.5).

Academic Background: B.A. in Meteorology from University of Northern Colorado

In Montana where Fransen lives and works, there are more cattle than people. According to the U.S. Department of Agriculture there are approximately 2.5 million head of cattle in Montana compared to only a million people. With livestock such a big part of the state's economy, losses due to extreme weather are a serious problem. The 2010 USDA livestock loss data for all livestock types shows nearly a billion dollars in losses due to weather around the United States. As of 2011 across the United States, cattle and calf

Figure 4.5 Tanja Fransen. *Source: Reproduced with permission from T. Fransen*

losses due to weather totaled a staggering 13% of all non-predator losses, or a total of 489,000 cattle/calves at a value of more than $274 million U.S. dollars (USDA National Agricultural Statistics Services Reports, 2010).

Ranchers reported that winter weather and cold events during birthing/calving season have the greatest economic impacts on their operations. Fransen worked with her NWS colleagues and the University of Miami to develop the Cold Advisory for Newborn Livestock, or CANL, to help mitigate economic impacts due to livestock losses. Newborn livestock refer to those less than 24 hours old. CANL was used extensively beginning in the winter of 2009, and nine other forecasting offices are utilizing it operationally as well. CANL uses wind chill, precipitation, and sky cover data to create an index that details the likelihood of a newborn calf experiencing hazardous weather conditions. Color-coded, the index ranges from green for "none" to red for "extreme" and is updated every six hours. The forecast goes out 36 hours. The CANL system lets ranchers know that there may be a deadly combination of precipitation and wind chill to endanger their calves, and sometimes it may not be enough to trigger a traditional warning or advisory. It provides a heads up so they can provide additional shelter or more feed (Fransen, 2010). Fransen said, "Our research and the discussions we had with the ranching community show the key elements they are worried about is wind chill and accumulating precipitation which can reduce the ability for the animal to dry off and warm up in the first crucial hours after being born." The ranchers also want to know if it will be sunny or cloudy, as sunny skies help dry off the newborn animals faster.

The index gives ranchers enough lead time to get their livestock to shelter and/or increase feed and water, reducing the possibility of losses. CANL provides ranchers with specific information they can use, filling in the gaps in existing NWS products. "While we have things such as winter storm and blizzard warnings, both obvious problems for ranchers, the CANL system hits those areas that the NWS would not generally warn for, such as light accumulating rainfall and wind and cold that doesn't meet normal thresholds," according to Valley County, Montana Sheriff Glen Meier, who is also a rancher, "It allows us to have at least 24 hours advance preparation in trying to mitigate losses of newborn livestock."

CANL is an essential first step toward helping livestock producers across the country to reduce winter weather-related losses. Fransen says, "Be open to new ideas, and see what can come from them. I never thought I would be an agency expert on weather and livestock, but I am now! This project had me look at the needs of the public first, not just creating something and then maybe evaluating it later to see its effectiveness. We got feedback first, developed a little, tested a little, got more feedback and worked on refining it until the users gave the final approval on the final results."

4.5 Fransen's Recommendations for Integrating Social Science and Atmospheric Science

1) Develop and nurture partnerships outside of your agency or academic circles. Involve people outside of your discipline area to bring in expertise. You may start a few projects with people that don't go far and you may not always know which ones will go the final mile, but taking the time to have conversations and discussions with others can eventually result in advances that benefit the public;

2) Don't expect things to happen overnight. Culture change takes time, but by providing consistent messages over and over, explaining the need for the change and leading by example, change(s) eventually happen;

3) Have passion for what you are doing, and believe in your work. Your partners and end users will appreciate it. If you are in over your head or just not passionate about a project, find someone to replace you who has the strengths to make it successful;

4) Be open to feedback, even if it is criticism. Getting thoughts and questions from people who think differently than you can improve what you are doing. It will make your work more successful in the long run if you involve others who have different opinions and ways of approaching things in the process; and 5) Seek other perspectives and be willing to learn from them. Social media offers a lot of opportunities to find groups that can introduce you to great resources. And network! You never know what the person you've just met may be able to help you.

Holthaus, Eric   sciencebyericholthaus@gmail.com
Holthaus writes a blog on weather and in 2016 he launched a podcast called "Warm Regards." He lives in Tucson, Arizona. In 2013, he vowed never to travel by plane again to reduce his carbon footprint (http://www.slate.com/authors.eric_holthaus.html) (Figure 4.6).
Academic Background: B.S. in Meteorology from St. Louis University and M.A. in Climate and Society from Columbia University (http://www.columbia.edu/cu/climatesociety/program.html)

Figure 4.6 Eric Holthaus. *Source:* E. Holthaus.

Holthaus is an applied meteorologist and journalist with 11 years' experience working on weather index insurance and climate impacts projects around the world. While working at Columbia University after completing his master's degree between 2008 and 2011, he served as a primary technical advisor and the field lead for weather index design and methodology for the HARITA weather insurance project in the Tigray region of Ethiopia, sponsored by Oxfam America, USAID, the World Food Programme, and Swiss Re, a global insurance company. Based on the products he co-designed, this project scaled up to other countries around the world. At Columbia University, he worked with Jeffrey Sachs as a lead technical advisor on UNDP/Columbia University's Millennium Villages Project where he designed weather indexes for 12 countries in Africa. He has also consulted on weather index insurance scoping projects in Uruguay, Central America, and Indonesia as part of the International Research Institute for Climate and Society at Columbia University.

Holthaus helped design disaster preparedness decision support systems making use of weather and climate forecasts, focusing on tropical cyclone risk in the Caribbean (particularly in Haiti where he worked with the Haitian Red Cross), the International Organization for Migration, and other humanitarian organizations. His team made arrangements for small business loans and drought insurance for farmers in Ethiopia. Between 2010 and 2013, he wrote a daily column focusing on weather and climate impacts in the Greater New York region called Weather Journal for the *Wall Street Journal* (wsj.com/weatherjournal). As of 2017, he writes about weather and climate for Slate's Future Tense. Follow him on Twitter: @EricHolthaus.

4.6   Holthaus' Recommendations for Integrating Social Science and Atmospheric Science

Holthaus says, "The weather is one of the few things people have in common throughout the world. It unites us. Those of us in the atmospheric sciences should start acting like it! Weather and climate don't have to be data-intensive, jargon-filled disciplines that exclude passionate researchers in the social sciences and the general public. There's so much common ground that's ready to be explored."

Figure 4.7 Heather Lazrus. *Source:* University Corporation for Atmospheric Research.

Lazrus, Heather   hlazrus@ucar.edu

Dr. Lazrus is an environmental anthropologist and a Project Scientist at NCAR in Boulder, Colorado (https://staff.ucar.edu/users/hlazrus) (Figure 4.7).

Academic Background: B.A. in Anthropology and in Religious Studies from Victoria University, and M.A. and Ph.D. in Anthropology from University of Washington

Dr. Lazrus uses the theories and methods in her anthropological toolkit to investigate the cultural mechanisms through which all weather and climate risks are perceived, experienced, and addressed. She focuses on the interface between extreme weather and climate change. Her research contributes to improving the utility of weather forecasts and warnings, reducing social vulnerability to atmospheric and related hazards, and understanding community and cultural adaptations to climate change.

Her dissertation, *Weathering the Waves: Climate Change, Politics, and Vulnerability in Tuvalu*, is an ethnographic examination of the political ecology of climate change impacts and the governance of vulnerability in a Pacific Island community (Lazrus, 2011). In 2015, she was engaged in research projects that surveyed public perceptions of and behavior regarding flash floods among residents of Boulder,

Colorado, hurricanes among vulnerable populations in Miami, Florida, and droughts among water users in south-central Oklahoma. Dr. Lazrus founded and serves on the American Anthropological Association Task Force on Climate Change.

She is dedicated to promoting meaningful social science engagement with weather and climate research and applications, including organizing a series of workshops at NCAR called Rising Voices of Indigenous Peoples in Weather and Climate Research and Policy. Participants come together to discuss what the science, information, support, and research needs are of tribal communities to facilitate respectful and appropriate adaptation solutions to climate change and variability. Rising Voices is a community of engaged Indigenous leaders, indigenous and non-Indigenous environmental experts, students and scientific professionals across the United States, including representatives from tribal, local, state, and federal resource management agencies, academia, tribal colleges, and research organizations.

Dr. Lazrus' interest in anthropology comes from her curiosity about how people create, learn, and share ideas, values, and behavior. As an undergraduate, she combined this developing interest in anthropology with her lifelong passion for environmental issues—likely the outcome of growing up in the mountains outside of Boulder, Colorado and then along the coast of New Zealand's North Island. She already recognized that environment and culture are intimately intertwined, and environmental issues are at their core about cultural values.

Dr. Lazrus found an Environmental Anthropology program at the University of Washington where she pursued graduate studies on the impacts of climate change in the Pacific Island country of Tuvalu. In the process of conducting ethnographic research, she met other researchers who inspired her to understand that weather and climate are on a continuum. She applies the same methods and theories grounded in anthropology that she used to understand climate change impacts in Tuvalu to examine the cultural mechanisms through which all weather and climate impacts are experienced, understood, and responded to.

4.7 Dr. Lazrus' Recommendations for Integrating Social Science and Atmospheric Science

Interdisciplinary collaboration must be based on the strong disciplinary foundations of all the collaborators. Collaborators bring their diverse disciplinary perspectives to bear on a compelling problem that couldn't be addressed by any one discipline alone. Interdisciplinary work requires passion for the problems being addressed, and it takes resources (especially, but not only, financial) to support the researchers who may encounter challenges working beyond the borders of their discipline, for example, publishing and tenure requirements.

Morss, Rebecca   Morss@ucar.edu

Dr. Morss is a Senior Scientist at NCAR in Boulder, Colorado and Section Head of the Mesoscale and Microscale Meteorology Laboratory (https://staff.ucar.edu/users/morss) (Figure 4.8).

Academic Background: B.A. in Chemistry from University of Chicago and Ph.D. in Atmospheric Science from the Massachusetts Institute of Technology

Dr. Morss has always been interested in combining science with societal relevance, and she has done this in various ways throughout her career. After receiving her Ph.D., she started formally studying societal aspects of weather as a post-doctoral researcher. Her current work focuses primarily on combining societal and meteorological perspectives to understand and improve communication, interpretation, and use of information about weather-related risks.

Dr. Morss has been a driving force behind integrating social sciences into the meteorological community, through her contributions to the WAS*IS movement and her research, mentoring, and community service efforts. This include co-authoring the National Research Council reports *Completing the Forecast*, which examines how to

Figure 4.8 Rebecca Morss. *Source:* University Corporation for Atmospheric Research.

improve the communication of uncertainty for better decision making using weather and climate forecasts, and *When Weather Matters*, which calls for side-by-side collaborations among meteorologists and social scientists to address challenges of urban meteorology and high impact weather (Committee on Estimating and Communicating Uncertainty in Weather and Climate Forecasts, 2006).

4.8   Dr. Morss' Recommendations for Integrating Social Science and Atmospheric Science

Dr. Morss offers five ways to integrate social science and atmospheric science: 1) follow in others' footsteps when you can, but forge your own path when you need to; 2) take feedback constructively and be open to new ideas and opportunities as you progress through your career; 3) seek out strong, critical mentors who will help you use your skills in the best way possible, and who will help you stretch your thinking and abilities to set and accomplish new integrated weather-society goals; 4) pursue your goals with the end in mind, but don't be afraid to incorporate new perspectives or take risks (even if it takes more time and effort); and 5) try not to worry too much about the twists and turns along the way, and strive for excellence in whichever paths you choose.

Moulton, Rebecca   rebecca.moulton@fema.dhs.gov
Moulton is a Hurricane Program Specialist with the Federal Emergency Management Agency in Atlanta, Georgia (Figure 4.9). Academic Background: B.S. in Communication from Ohio University and M.S. in Earth and Atmospheric Science from Georgia Institute of Technology

Like most meteorologists, Moulton knew what she wanted to do at an early age. And like many others before her, her lifelong dream of becoming a meteorologist was (nearly) dashed on the cliffs of calculus and differential equations. As an undergraduate in the early 1990s, there were still relatively few meteorology programs and even fewer mentors in the field. As she reached out for advice, she was told that pursuing a career in meteorology is too competitive with very few job opportunities. Given her skills and interest in communication, perhaps she would be better suited in broadcast meteorology or a different field altogether.

Moulton graduated with a communication degree, unsure where that would take her. Everyone told her she was a "people person," so

Figure 4.9 Rebecca Moulton. *Source:* Reproduced with permission from R. Moulton

she went into sales. She hated her sales job and only lasted three months. After a few other jobs and much soul searching, she decided to go back to school to pursue her true passion—weather! This was no easy task. It required her to find a mentor who believed in her potential and to work harder than she had ever worked before to earn her master's degree in Earth and Atmospheric Science from Georgia Tech. She began working at the Weather Channel when she was a graduate student and continued when she graduated. During the active 2005 hurricane season, she worked as a liaison with emergency managers. Soon, she began hearing others in leadership recommend that meteorology students consider taking courses in communication, and she realized the potential of her experience and backgrounds.

Moulton soon joined FEMA as a Hurricane Program Specialist where she uses both the communication and meteorology sides of her background in education and outreach, training, planning, and operational response to hurricanes. She works closely with the National Hurricane Center to help convey the threat to emergency managers, and she works with emergency managers to develop evacuation zones and plans that incorporate public behavioral response. "I always saw communication background as a liability... but it has turned out to be the foundation of my career in meteorology and emergency management."

4.9 Moulton's Recommendations for Integrating Social Science and Atmospheric Science

Reach out to others. Build a network of like-minded, hardworking experts in various fields to use as resources and COLLABORATE! Don't try to forge this alone—go at it as a team. Through WAS*IS, my professional colleagues, associations, and other groups, I have a network of experts in various fields I know who I can "go to" for expertise, advice, and experience in a variety of areas. Having a cadre of like-minded meteorologists who I know are interested who can help champion the way forward is a valuable resource—and inspires me daily when I run into challenges, obstacles, and new problems.

Ruin, Isabelle   isabelle.ruin@ujf-grenoble.fr

Dr. Ruin is a CNRS (French National Centre for Scientific Research) Research Scientist at LTHE (Laboratoire d'étude des Transferts en Hydrologie et Environnement) in Grenoble, France (http://www. lthe.fr/pageperso/ruin/Isabelle_Ruin/Welcome.html) (Figure 4.10). Academic Background: M.Sc. in Earth Sciences and Geotechnics at University of Clermont-Ferrand, M.A. in Geography and a Ph.D. in Geography from University of Grenoble

Dr. Ruin's early passion for earth sciences and especially for volcanic activity brought her to study geology and volcanology where she

Figure 4.10 Isabelle Ruin. *Source:* Reproduced with permission from I. Ruin.

specialized in geotechnics applied to natural hazards. After obtaining a master's degree in Earth sciences, she worked in non-profit organizations dedicated to environmental education and local sustainable development. Together with educating children about the richness and fragility of their territory, she was in charge of a project involving volcanologists and elected officials for the development of ecotourism in a volcanic and rural landscape. This experience made her "realize that Nature can't be considered without taking into account the Human component and conversely." She decided that studying Geography was her best option, and that the specialization on natural risks offered the perfect setting for coupling human and natural system considerations.

A master's degree in geography was an opportunity to refine her project and find the ideal advisors in both human and natural sciences for pursuing a Ph.D. After completing a doctorate dedicated to studying human vulnerability and especially the susceptibility of motorists to flash flood events in France, she spent two years as an Advanced Study Program postdoctoral fellow at NCAR. She received a grant from the French National Research Agency and went back to France in late 2009 to join the LTHE research lab as the principal investigator of a three-year project called ADAPTflood.

Dr. Ruin is a geographer at CNRS who works daily with physical and social scientists on hazard reduction research, primarily related to flash floods. She is fluent in the vocabularies, methods and concepts of both physical and social science. She works on the campus of Grenoble Alpes University in a supportive environment. She works with an interdisciplinary team at LTHE (a research lab focused on the hydrologic cycle and its links with the Earth's climatic and environmental changes) and in close relation with the PACTE (a research lab composed of geographers and political scientists). By learning how people, particularly motorists, are exposed to flooding conditions, how they react to warnings and adapt their habits and daily behaviors in heavy rains and flash flood conditions, they hope to help develop warnings that alter those dangerous behaviors and reduce flood losses.

Dr. Ruin and her colleagues are integrating space and time scales of both the physical phenomenon and human activities to understand how individuals anticipate or cope with dangerous conditions with very little lead time. Ruin's work was originally funded by an insurance company. Her research is currently funded by the French National Research Council (ANR) and stands on the frontier between basic and applied research. It aims at targeting decision makers to provide them locally and culturally relevant options based on the lessons learned from the observation of resilient practices. She is also publishing research articles on flash flood time and spatial scale with her Ph.D. students (e.g., Terti *et al.*, 2015).

4.10   Dr. Ruin's Recommendations for Integrating Social Science and Atmospheric Science

I learned three main take-home points from my own experience that are important for realizing true integration: 1) find the right people to talk to: Not only do you need to be curious and open to other disciplines and approaches, but you need also to find experts in other fields who are willing to listen and consider your thoughts and ideas, as much as you do to theirs. Finding collaborators is crucial in building interdisciplinary projects; 2) experiment in the field: meet and talk with "real" people face-to-face, including officials who have local knowledge of end-users. Also, share field experience (surveys, methods, experiment protocols) with experts of other disciplines who will learn a lot from your own work and; 3) build sustainable collaborations: Interdisciplinary work is not straight forward, it takes time and energy and needs to be thought of on a long term basis, building strong foundations step by step. I believe

Figure 4.11 Russ Schumacher. *Source:* Reproduced with permission from R. Schumacher.

that targeting sustainable and steady collaborations is more rewarding than aiming at conjectural opportunities.

Schumacher, Russ   russ.schumacher@colostate.edu

Dr. Schumacher is an Associate Professor in the Department of Atmospheric Science at Colorado State University, Fort Collins, Colorado (http://www.atmos.colostate.edu/faculty/schumacher.php) (Figure 4.11). Academic Background: B.S. in Meteorology and Humanities from Valparaiso University in 2001, and M.S. (2003) and Ph.D. (2008) in Atmospheric Science from Colorado State University

Before landing a position at his alma mater Colorado State University, Dr. Schumacher was an Advanced Study Program Postdoctoral Fellow at NCAR from 2008 to 2009 and an assistant professor in the Department of Atmospheric Sciences at Texas A&M University from 2009 to 2011. Schumacher received a prestigious NSF Career Grant for five years of funding. He developed the SPREAD program as part of his work. SPREAD was a two-part workshop that brought together hydrology, meteorology, history, economics, and sociology PhD students for a week in 2013 and a week in 2014 to develop improved ways to predict heavy rainfall that lead to flooding and ways to communicate that information more effectively. Further details can be found in Schumacher 2016, listed in references.

4.11   Dr. Schumacher's Recommendations for Integrating Social Science and Atmospheric Science

We are getting to a point where weather forecasts are routinely so good that the primary challenge for meteorologists is presenting and communicating those forecasts to the people who want and need them. But there is one big exception to this: truly extreme, high-impact weather events, which in many cases remain the most challenging to predict. And since we can't predict these truly impactful phenomena with 100% certainty, we need to make probabilistic forecasts of them. And this increases the challenge of presenting and communicating the forecast even more: percentages, odds, and likelihoods can be difficult to understand. So this leaves us with a couple of major problems that we need to tackle going forward: 1) improving predictions of high-impact weather, which will require better observations and models of the atmosphere, and better understanding of how they work; and 2) improving the way we communicate those forecasts, which will require meteorologists and those with expertise in many other areas to work together so that the improved forecasts also equate to the things that really matter: lives saved, property losses reduced, and economic stability restored.

Dr. Schumacher is part of the group of early career scientists that makes advances in both of these areas, as an educator and a researcher. He wants to make sure his students have an appreciation for not only the scientific and technical aspects of the atmosphere, but also for how the atmosphere impacts people and how our science might help to reduce the negative impacts. Dr. Schumacher hopes to do research that leads not only to better knowledge of how to understand and predict the weather, but to how people can use that understanding and those predictions.

Spinney, Jennifer   jen@jenspinney.ca

Spinney is a research anthropologist who conducted an ethnographic study of emergency managers to help weather forecasting software developers incorporate the expressed needs of emergency managers into new weather forecasting software (Figure 4.12).

Academic Background: B.S. in Exercise and Sport Science from University of Manitoba, and M.A. in Anthropology from University of Western Ontario. In 2017, she will complete her Ph.D. in Anthropology from University of Western Ontario.

Figure 4.12 Jennifer Spinney. *Source:* Reproduced with permission from J. Spinney.

Using in-depth interviews in flash flood alley in Texas including Austin, New Braunfels and San Antonio and in the greater Kansas City metropolitan area, Spinney has helped forecasting software developers at the NOAA Global Systems Division Integrated Hazard Information Services (IHIS) incorporate the emergency managers' expressed needs into new forecasting software. The project part was of the Integrated Hazard Information Services (IHIS) effort and an innovative 2010 to 2012 partnership between social scientists and software developers.

Spinney's master's thesis "Perceptions of Vulnerability to 'Severe' Weather in Pangnirtung, Nunavut" compared how the officials and the locals perceived the impacts of a bad storm (Spinney and Pennesi, 2012). Spinney was the invited social scientist on the 2011 Joplin, Missouri National Weather Service Regional Survey Assessment Team (www.nws.noaa.gov/os/assessments/pdfs/Joplin_tornado.pdf).

Spinney's doctoral project is an ethnographic investigation of urban and flash flood warning and response in southwestern Ontario. It is a collaborative undertaking with Environment Canada (EC) and Conservation Authority (CA) forecasters. Through participant observation in forecast offices, semi-structured interviews with forecasters and the public, and the distribution of a public survey after a flood event, the results of the study will be used to enhance EC and CA flood warning production and communication. The results will also inform theories on multi-level governance, the production of scientific expertise, risk knowledge, risk communication, as well as sense making and decision making.

4.12   Spinney's Recommendations for Integrating Social Science and Atmospheric Science

Spinney offers four recommendations: 1) have social scientists as collaborators and/or partners from the very beginning of projects, perhaps as co-leads and contributors during the development of project objectives; 2) ensure funding is available for social scientist collaboration for the entire duration of projects (just as it is for physical scientists working as part of teams), and not something that needs to be renegotiated or re-proposed or applied for every year; 3) embed social scientists within teams instead of relying on consultants to carry out work thereby making social scientists part of the project and not simply adjuncts or ad hoc members; and, 4) develop software and official NWS products based on users' needs, definitions of risk, and risk thresholds.

Figure 4.13 Lisa Vitols. *Source:* Reproduced with permission from L. Vitols.

Vitols, Lisa Lisa.Vitols@ec.gc.ca

Vitols is the Engagement and Strategy Advisor for the Meteorological Service, Environment Canada (Figure 4.13).

Academic Background: B.A. in History and Geography at McGill University and M.A. in Environmental Education and Communications at Royal Roads University

Vitols' government career began in Ottawa working as a political assistant to federal Ministers in several government departments, ending in the environment department. As a public servant, she has worked in a variety of divisions, including Intergovernmental Affairs, Environmental Policy, Communications, and now the Meteorological Service of Canada (MSC). Her qualitative master's work and her "non-meteorologically tainted" perspective is what make her an asset working with the MSC. Vitols says: "I try to help the MSC better understand Canadians, and Canadians to better understand the MSC. During the 50 preceding years before I was hired, the department never asked Canadians what information they wanted from the weather service – they were just given whatever the service felt like giving, or what they assumed people would want. We now recognize that as a government and a service, we should reflect citizens' needs better, and to do that, we need to have a better social science perspective on what that is – it's not just a matter of how many hits on a particular webpage. We need to listen to, collect, and re-tell stories from real people so that information can be best provided for their most effective decision-making."

Interesting projects for Vitols have included a "wind perception study," asking Canadians on the street how fast they thought the wind speed was and how it could best be portrayed graphically; "Let's Talk Weather," an online engagement website where questions were posed to gather feedback and citizen's weather stories collected; and the evaluation of various weather-related pilot projects related to the 2010 Olympic Winter Games.

To integrate social and atmospheric science, both raising awareness and proving the usefulness of that integration are key. It helps to have champions of the integrated approach along with qualified and connected people who understand the science but who don't need to be physical scientists themselves. MSC has found benefit from having multiple perspectives, working with people from a mix of backgrounds, and using external feedback, ever more qualitative in nature, to validate their work.

4.13 Vitols' Recommendations for Integrating Social Science and Atmospheric Science

Vitols says, "Network, make connections; stick your nose in where it should rightfully belong, try to get in on it before something starts, educate as you go along." Vitols offers three examples of how MCS is integrating social science and meteorology:

1) we worked closely with a national amateur sport organization to provide lightning awareness and wind chill awareness messaging, pocket-size cards, and stickers to this highly weather-impacted audience;
2) we share our raw data feed with university researchers who turned it into a mobile app that gives UV level notifications and impacts of UV exposure (http://uvcanada.ca/); and,
3) we meet with decision makers across the country (e.g., municipal and provincial emergency management officials, school districts, and chambers of commerce) to determine what their most important weather information for decision making is—this helps inform our product and service development.

4.14   Questions for Review and Discussion

1   The profiles include early and mid-career leaders who are working at the intersection of weather and society. What six U.S. federal agencies are represented in their backgrounds? Name four other government agencies at the local, regional, state, federal, or international levels that are not mentioned in the profiles but that are relevant to reducing losses from severe weather.

2   What struck you most about the variety of people profiled in the chapter? Do you identify with any of them in terms of directions you imagine your career might take? Why? Notice the evolution of careers morphing from broadcaster to NWS forecaster in Angle's case or from geologist to social science researcher in Ruin's case? Are you more attracted to a career that is more theoretical and research oriented or are you drawn more to the operational career or applied career in emergency management or hazard risk reduction? If you were to pursue questions on the intersection of weather and society, what area interests you?

3   What are the common themes found in the recommendations for integrating social science into atmospheric science? Do the recommendations focus more on academic or scholarly activities or more on collaborative or partnership activities? Are there specific steps you can take to encourage better cross-disciplinary or interactive work between you and your colleagues?

4.15   Using What You've Learned: Homework Assignment From the Chapter

1   Select two people whose weather and society work you have admired. It can be people from your readings, your university, your work, your social media connections, or somewhere else. Think broadly to include disaster mitigation, climate change adaptation, or others if you are interested in these topics. Select people with different roles such as a research scientist and a social media blogger. Do some web research related to the background and earlier career moves of those people. Write three paragraphs

detailing what you found and discuss what surprised you about how they got to their current job. How do people in different careers represent themselves differently on the web?

2 Find or follow a web link to one of the people profiled in the chapter. Since the book's publication, in what new endeavors are they engaged? Write two paragraphs identifying whom you selected and summarize their most recent accomplishments in terms of publications, new career moves, or public outreach.

References

Committee on Estimating and Communicating Uncertainty in Weather and Climate Forecasts. (2006) https://books.google.com/books?hl= en&lr=&id=RnPE4bXzk50C&oi=fnd&pg=PP1&dq=committee+on+ Estimating+and+Communicating+Uncertainty+in+Weather+and+ Climate+Forecasts,+2006&ots=6cHVIx6DhC&sig=nVQa5X30ffi9vK2 FqjD1tM1-I6w#v=onepage&q=committee%20on%20Estimating%20 and%20Communicating%20Uncertainty%20in%20Weather%20 and%20Climate%20Forecasts%2C%202006&f=false (accessed July 23, 2017).

Dobson, G., and Boehnert, J. (2015) National Center for Atmospheric Research. *GIS Tutorial for Atmospheric Sciences*, http://gis.ucar.edu/ projects/course-introduction-gis (accessed July 23, 2017).

Fransen, T. (2010) Cold advisory for newborn livestock: The stars aligned. Weather and Society Watch 2. http://www.sip.ucar.edu/news/ volume4/number2/focus2.php (accessed July 23, 2017).

Lazrus, H. (2011) *Weathering the waves: Climate change, politics and vulnerability in Tuvalu*. New York: Proquest.

Spinney, J.A., and Pennesi, K.E. (2012) When the river started underneath the land: social constructions of a 'severe' weather event in Pangnirtung, Nunavut. *Canada Polar Record*, 1–11.

Terti, G., Ruin, I., Anquetin, S., and Gourley, J. (2015) Dynamic vulnerability factors for impact-based flash flood prediction. *Natural Hazards*, 79(3): 1481–1497.

USDA National Agricultural Statistics Services Report. (2010) https:// www.nass.usda.gov/Publications/Ag_Statistics/2010/ (accessed July 23, 2017).

5

Moving Toward Integrated Weather and Society Research and Practice—A New Paradigm

"Too often we have overlooked the interdependency of physical science and social science and how they complement each other. The physical sciences help us make our forecasts and warnings more accurate. However, no matter how accurate our forecasts and warnings are, if we fail to convey risk in a manner that creates appropriate action, our mission is incomplete..." Ken Graham, Meteorologist-in-Charge, National Weather Service, Slidell, Louisiana (after 2005 Hurricane Katrina).

5.1 How Social Scientists and Meteorologists Work Together to Create New Scientific Conceptual Models and Methods: Start with Adjacent Projects

The first step in the move toward working with people with different backgrounds is often to formalize arrangements for atmospheric scientists and a variety of social scientists to work on collaborative projects. The scientists work next to each other (either in person or through a shared platform) on a common problem. The interaction is part of building trust and common language for more integrated collaborations in the future. Integrated work is a more advanced process and hopefully can follow "adjacent" projects in many cases.

A common problem with adjacent work is bridging the differing expectations for what the research will produce and how it can be used. The main advantage of the adjacent efforts is that the partners must get to know each other professionally. They have opportunities

Weather and Society: Toward Integrated Approaches, First Edition. Eve Gruntfest.
© 2018 John Wiley & Sons Ltd. Published 2018 by John Wiley & Sons Ltd.

to learn about each other's contexts, constraints, and concerns, and the first step can lead to more productive integrated research.

Adjacent work can be frustrating as well as exhilarating. Sometimes these collaborations are called multidisciplinary or interdisciplinary. Often the partners have different or unrealistic expectations of what's possible and how long it will take. Also, there are numerous pragmatic challenges of integrated research. It is not simple to connect datasets. Initial contacts can be very frustrating because professionals come with different priorities and languages. The following three case studies of integrated weather and social science research projects illuminate the challenges and rewards of creating this new paradigm.

Case Study: Schools and Tornado Saferooms

Should schools in "tornado alley" be required to build safe rooms? If so, who should pay for them? Are schools safer than the other alternatives of students being caught outside or in mobile homes? Should school districts without in-school shelters have tornado days like they have snow days where they close for the day? Do the school districts base their decision to close school due to tornados as a way to shift liability or because of their primary interest in student's safety? How do we find answers to these questions? Who should provide input?

After public school students die in tornadoes in schools, as seven students did in Moore, Oklahoma in 2013, debates heat up about where students are safer and what the school districts' best options are for reducing loss of life in tornadoes. A major component for such decisions is cost. Tornado shelters are expensive. Economist Dr. Kevin Simmons has taken a stance on his blog: http://kevinmsimmons. blogspot.com/. He considers public versus private expenditures for tornado mitigation activities. He reiterated his research finding that using public money to save lives from tornado fatalities "fails the benchmark that spending is considered reasonable if the cost per avoided fatality is less than $10 million." In their book, *Economic and Societal Impacts of Tornadoes and the Deadly Season: An Analysis of the 2011 Tornado Outbreak*, Sutter and Simmons show that these programs are well outside that benchmark for most uses of the program. Using public funds to reduce fatalities in mobile homes is the one exception for some states. Simmons has studied the economics of building codes and has published widely and also has a blog where he writes about these matters (http://kevinmsimmons.blogspot.com/).

Superintendents of public schools are weighing in on the topic, too. The decision of whether or not to build tornado shelters for schools also depends on whose funding will be spent on the shelters. Will the money come from FEMA or somewhere else in the federal government, from the state or from the city or school district? Monte Thompson of Wagoner, Oklahoma said: "There are essentially three ways to go about securing funding for shelters. The federal government FEMA will pay 75 percent of a constructed shelter and the district or city would be responsible for 25 percent. Other options are for the state or local governments or a consortium of private companies in cooperation with governments to pay for the shelter" (http://www.ktul.com/story/22462255/storm-shelters-in-schools-wagoner-super intendent-says-safety-is-top-priority, written by Caitlin Alexander, June 6, 2013).

What about having "tornado days" similar to "snow days?" Between 2005 and 2011, most of the tornadoes occurred between 2:30 p.m. and 7:30 p.m. (Wertz, 2013). After the 2013 Moore, Oklahoma tornadoes, the Tulsa, Oklahoma School District emergency management coordinator announced that he would not call tornado days. He mentioned concerns about canceling school on days when no storms occur and dismissing students early that can lead to students being stranded, alone, or stuck in traffic at the time when the tornadoes arrive (http://stateimpact.npr.org/oklahoma/2013/07/11/some-shelterless-oklahoma-schools-to-cancel-class-when-tornados-threaten/).

In Enterprise, Alabama after an EF-4 tornado severely damaged the high school on March 1, 2007, eight students were killed at the school. If the school system in Enterprise, Alabama were to have let the kids out sooner, those kids would have been at home.

Following the May 2013 tornadoes in Oklahoma, some North Alabama school districts decided they would close when the NWS SPC issued a tornado watch. Schools were closed at mid-day on October 26, 2010 (Cumbow, 2010). The schools dismissed the students between 11:30 a.m. and 1:15 p.m. The tornadoes were expected between 11 a.m. and 8 p.m. depending on location in Alabama.

NOAA Research Meteorologist Dr. Harold Brooks said schools are fairly safe places, in the bigger context. He said, "Just 13 children have died at school during a tornado in Oklahoma history, including the seven in 2013. There have probably been more people killed going to and from school than there have been tornadoes" (Ogle, 2013). Brooks writes that having a purpose-built shelter is ideal, but not everyone has one. The message for people who are under a NWS tornado warning is to "get as low as you can and put as many walls as you can between you and the tornado." This can mean go to the lowest floor, interior room, bathroom, or closet (Brooks, 2013). Oklahoma was considering school closures for tornado warnings too, similar to snow days (Layton, 2013).

Public school officials have difficult choices to make when confronted with severe weather forecasts. Tornadoes pose particularly difficult decision challenges because the tornado impacts can be deadly and the forecasts always come with some degree of geographic uncertainty.

Case Study: Storm Surge Warnings From the NWS—Moving From Adjacent Social Scientific Collaboration to Integrative Collaborations

The NWS invited social scientists from NCAR and elsewhere to help develop new materials to improve public understanding of storm survey forecasts (http://www.stormsurge.noaa.gov/r_and_d.html). New experimental product prototypes for warnings for storm surge were presented to emergency managers, broadcast meteorologists, and the public for thorough evaluation before eventual implementation. Trying out different wording options and map graphics with various groups allowed NWS to assess how to communicate storm surge information in ways that promote appropriate actions in vulnerable areas. Morrow *et al.* evaluated the social scientist and NWS collaboration (2015).

The social scientists found through conducting interviews and surveys with emergency managers and other stakeholder from local and state governments in coastal areas that a significant portion of the surge-vulnerable population does not understand what storm surge is (including its causes and how it works), what their vulnerability is, and what the potential impacts are. The study respondents also said they, too, wanted more information from NOAA's storm surge experts.

An important component of the research process was to allow stakeholders to raise issues affecting the communication of surge risk that were not necessarily contained within the closed-ended questions. Several issues were identified in open-ended comments in the various survey efforts. A common concern was the timing of storm surge forecasts. Emergency Manager concerns included the following: "Releasing storm surge forecasts 2 days before landfall is practically useless. Evacuation orders would have been initiated by that time. There are private services that are already providing that information" (Morrow and Lazo, 2013a:17); "The NWS doesn't seem to recognize the reality of modern day, 24 hours a day media coverage of these events" (Morrow and Lazo, 2013a:17); and "The NWS needs to issue storm surge predictions way earlier. I think they are afraid they'll be wrong, but by the time they issue the official forecast, it is largely irrelevant" (Morrow and Lazo, 2013a:17).

Broadcasters expressed a "timing" need for the forecasts. They asked to receive them at least 15 to 20 min before the hour since their broadcasts usually begin on the hour. These statements point to the need for ongoing discussions between NWS, broadcasters, and EMs as to the timing of information as well as the associated reliability.

The storm surge inundation graphic was used on an experimental basis in 2014. In 2016, the new storm surge products were operational for Tropical Storm Hermine (National Hurricane Center, 2016). Rarely do research results from stakeholders inform policy in such a timely manner. This social science research sets an important example, particularly in an era of diminishing resources, by combining funds across several NOAA offices to address critical questions. It also serves as an example of combined qualitative and quantitative research methods to explore issues and develop options that can then be tested empirically (Morrow *et al.*, 2015:44). The Morrow team's storm surge social science research represents a significant effort on the part of NOAA to empirically elicit stakeholder input as part of product development. It brought in the perspectives of relevant stakeholders, including emergency managers, broadcasters, and the general public, rather than relying only on input from self-selecting respondents.

The initial social science results showed a disconnection between NWS's perception of storm surge risk and the public perceptions of the problem. The proposed new NWS storm surge "products" did not address the concerns that the vulnerable population had or the time frames used by emergency managers. This mismatch could be a result of poor timing. To be most useful, the social science research has to occur before the development of the NWS warning product, so that the new products can best reflect the expressed needs and understanding of the stakeholders.

By the time the NWS invited the social scientists to participate in the product development process, NWS was already significantly invested in the form and wording of the storm surge warning tools they wanted to use. This was problematic. The lesson learned is that social scientists should be engaged by an agency as true product development collaborators from the beginning, not as marketers at the end of development.

The collaborations between the social scientists have evolved and improved. The result has been more interactive and productive (Lazo *et al.*, 2015; Morrow *et al.*, 2015; Morrow, 2013).

Case Study: Required Integrated Research: National Science Foundation Dynamics of Coupled Natural and Human Systems Program

The National Science Foundation (NSF), NOAA, NASA, and many other funding agencies have research programs that set aside funding for collaborative projects across physical and social sciences. Their scope goes beyond weather and society and covers many ecological, biological, health-related, and other collaborations. The NSF Dynamics of Coupled Natural and Human Systems program is an example of a program that requires that the researchers propose projects that go beyond "adjacent" research and address problems that require integrated approaches (http://www.nsf.gov/funding/pgm_summ.jsp?pims_id=13681). Integrated approaches require changes to the ways that the intersection between atmospheric science and social science are considered. NSF and other agencies are encouraging these changes.

5.2 Increased Popularity for How Important It Is for Meteorologists to Understand Some Social Science

Dr. Marshall Shepherd is an atmospheric scientist who is an active champion for expanding the meteorological imagination beyond the atmosphere. He is a distinguished professor of atmospheric science at the University of Georgia. Since 2014, he has hosted The Weather Channel's Sunday talk show *Weather Geeks*, a pioneering Sunday talk show on national television dedicated to science. He also is a contributor to *Forbes Magazine Online*. His shows and articles bring together partners working on many aspects of weather and climate challenges.

In his academic work, TV shows, public appearances, and columns, he directly confronts key timely challenges at the intersection of weather and society. For example, in April 2016 following devastating floods in Houston, Texas, his *Forbes* column asked readers to explain why so many drivers in Houston had to be rescued from rising waters. There were more than 1200 high water rescues, more than 17 inches of rain and over 1000 homes flooded with between 1.3 and 1.9 billion dollars in damages (Yan and Mier, 2016). Shepherd aptly sums up many contemporary severe weather impact situations: "Like many things we see in meteorology today, the forecasts are really good. The life-threatening decisions are often rooted in decisions, communication, and perception" (Shepherd, 2016), suggesting using an integrated approach to answering his question about why people drive into high water. Shepherd shares his thoughts and suggests links to the WAS* IS Facebook page a few times a week (Figures 5.1 and 5.2).

5.3 Possible New Common Ground for Integrated Approaches to Weather and Society: Emergence of Convergence Science

Since 2000, NSF and the National Academy of Science have been developing "Convergence Science" (Roco and Bainbridge, 2003). Convergence has the potential to change how universities are organized and how disciplines are structured including meteorology and

Figure 5.1 Dr. Marshall Shepherd, Professor of Atmospheric Science, University of Georgia. *Source:* Weather Geeks. Reproduced with permission from The Weather Channel.

Figure 5.2 Dr. Marshall Shepherd on the set of his Sunday afternoon show Weather Geeks (WxGeeks) interviewing Dr. Kim Klockow and Dr. Susan Jasko at the American Meteorological Society Meeting in Phoenix, January 2015. *Source:* NOAA.

the social sciences. It offers new ways of discovery, innovation, and application opportunities through specific theories, principles, and methods to be implemented in research, education, production, and other societal activities. "Using a holistic approach with shared goals, convergence seeks to transcend existing human limitations to achieve improved conditions for work, learning, aging, physical, and cognitive wellness...The formulation of relevant theories, principles, and methods aims at establishing the convergence science. Convergence is a transformation model in the evolution of science and technology (S&T) that unites S&T fields with society. It provides a framework and approach for advancing not only science and engineering but also business and policies. Convergence is a deep integration of knowledge, tools, and all relevant areas of human activity to allow society to answer new questions, to create new competencies and technologies, and overall to change the respective physical or social ecosystems. Such changes in the ecosystems open new trends, pathways, and opportunities in the following divergent phase of this evolutive process" (Bainbridge and Roco, 2016:211).

An international group of scientists and practitioners developed the concept of "convergence" as a way to redefine the existing paradigm to be more problem-driven and to require more than one discipline's perspective. They aimed to bring together discipline, people, and ideas to work productively on very complex problems to generate ideas and capabilities not present in any one discipline. They showed how materials science and nano-science are examples of brand new sciences that have emerged from fundamentally new collaborations. The convergence group hosted workshops in Latin America, the European Union, Israel, Korea, Japan, China, and Australia from 2011 to 2012. The NSF team gathered constructive comments and hosted numerous discussions about unifying holistic approaches that are necessary to understand how society is integrated with nature. Convergence science will be driven by societal values and needs rather than by technologies and new products. In 2012 NSF, NIH, NASA, EPA, DoD, and USDA conducted a global study of convergence knowledge and technology, including converging platforms, methods of convergence, societal implications, and governance. The multi-agency work considered all science, including social and atmospheric sciences.

A brand new field that emerges from socio-eco-hydro-meteorology can transform the ways that forecasts are understood and used for

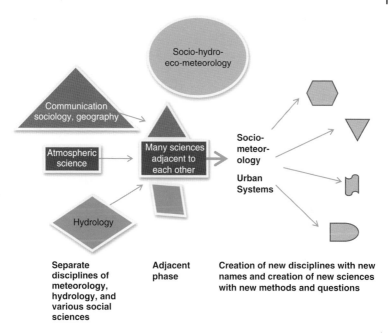

Figure 5.3 Adapted from the model of how science changes from disparate silos to adjacent or integrated entities to completely new kinds of science that then spin off new ways of understanding and completely new types of science. Roco termed the phases: creative, integration, and innovation. *Source:* Adapted from Roco *et al.*, 2013:141 with permission from Springer.

significant loss reduction. The new field would have its own name, theories, and methods. When the meteorological and social science communities do truly integrated work, new research questions would be addressed in new ways that positively increase resilience, reduce losses, and create new fields of study and practice. Figure 5.3 shows how the disparate sciences move from their individual perspectives, working on their own, to working adjacent or integrated with each other. The availability and quest to utilize big data for problem solving can accelerate the development of a new field beyond the traditional stove-piped disciplines within physical and social science (Holling, 2001).

Shared methodologies, theories, approaches, and common constraints, such as human values and sustainable development, are part of all new efforts. The new approach supports activities that are driven by *people and their needs*. Dr. George M. Whitesides is professor of chemistry at Harvard University and leads a multi-disciplinary group at Harvard that studies nanotechnology, society, robots, health, and many other contemporary societal and scientific problems (https://gmwgroup.harvard.edu/index.php). He comments that, "With tools we have now, we cannot solve large social problems. Our basic structure does not allow us to deal with the most important problems that we face." Beyond weaving the silos together, new hybrid ways of thinking and fusions are required. As Whitesides points out, "'...Convergence' changes the approaches used to solve scientific problems rather than letting technology or disciplinary specializations determine and drive which problems are addressed." Whitesides and his colleagues recognize that incentives will be required to get physicists to work with sociologists because present practices discourage truly interdisciplinary problem-solving (Whitesides, 2014). Many institutions, academic departments, and individuals thrive in the disciplinary stovepipes or silos as they are set up. They are content with the ways things are now. These "placeholders" sensibly and rationally obstruct possible changes in practice and thinking because it would upset their stable working environment and involve more work and conflict. However, such obstruction begs the question: Is it strategic to stand still as a hedge against losing ground in the movement toward integration between social sciences and atmospheric sciences?

The problems at the intersection of weather and society demand new integrated scientific approaches to get beyond incremental change. Tornadoes are one of the many types of hazards that people face where the best ways to reduce losses involve the intersection of physical and social science. Adding a few minutes of additional lead-time for the formal NWS warning for tornadoes is not the scientific leap that is going to make the most significant difference in the number of people who survive them. The reductions in losses will result from increased time combined with changes in the ways people behave when faced with tornado threats and changes in the options people have for reducing their own vulnerability. As long as people stay in buildings that are not structurally sound to withstand tornadoes, even relatively small events will kill people. Without the means to get to a stronger structure, some people will die. The combination of social science and physical sciences is key to seeing major improvements in reducing losses from weather events.

The first steps have involved social scientists and meteorologists working side-by-side on problem solving, participating in workshops like the Integrated Warning Team meetings where the media, emergency managers, forecasters, members of the public, and others address local or regional concerns, and working on projects that bring them together on spatial and temporal scales where they get to know each other. The Living with Extremes meetings also brought scientists from many disciplines together.

Advancing from these efforts will be meaningful but it will be incremental. For example, the color scheme for probabilistic warnings for tornadoes may be clearer and unambiguous, but will the information that is conveyed meet the needs of emergency managers, other decision makers, and the public at risk? We don't know, yet.

True progress requires new ways of thinking. A new discipline such as socio-hydro-eco-meteorology, hopefully with a new name, will emerge with its own sets of methods and the problem definitions that are formed with different perspectives than exist within the current disciplinary frameworks. Meteorologists can build upon the growing number of interdisciplinary communities. Even when all the partners enter the collaborative project with good intentions, difficulties can occur.

5.4   Socio-Meteorological Testbeds

How can new communities of scholars and practice be established where people from many disciplines talk among themselves and with broader communities? In these testbeds, anthropologists, sociologists, geographers, psychologists, and other social scientists must work not only with meteorologists, but also with ecologists, hydrologists, a wide range of practitioners, and many others. The scientists and practitioners will be co-producers of the new science and knowledge. Changing the way science considers weather and society (or changing its paradigm) will be based on concrete measures of success, not just "we talked to a few people to see how they liked it." Where do the new theories and practices hold up or fall apart in the real world? The socio-bio-eco-meteo-hydro sectors need to work together as all these factors make up the environment in which everyone lives.

Starting with stakeholder preferences for spatial information and temporal information rather than with the latest tool or product would lead to the development of new tools and products that are based on stakeholder needs and not on what is the latest and considered "best" technology. If the stakeholder views were not considered from the first consideration of the next tool or product, it may not truly improve stakeholder decisions and thus reduce losses. Social scientists and well developed social science research is necessary to meet stakeholder needs, whether the stakeholders are forecasters, emergency managers, tourists, parents, teachers, mayors, or others (Webster, 2016).

5.5 Wicked Problems and Momentum In the Movement Toward Integrated Weather-Society Science

In the 1970s, Rittel and Webber (1973) coined the "wicked problems" term in the context of problems of social policy to apply to problems for which a purely physical scientific-rational approach cannot be applied. They pointed out that the early challenges of the late nineteenth and early twentieth century were solvable. For example, cities built sewer systems underground and pumped water into people's homes to avoid water-borne diseases.

Many twenty-first century problems are qualitatively different. They lack clear problem definition, and the many stakeholders have different perspectives. Rittel and Webber wrote, "The search for scientific bases for confronting problems of social policy is bound to fail because of the nature of these problems...Policy problems cannot be definitively described...In a pluralistic society, there is nothing like the indisputable public good; there is no objective definition of equity; policies that respond to social problems cannot be meaningfully correct or false; and it makes no sense to talk about 'optimal solutions' to these problems...Even worse, there are no solutions in the sense of definitive answers" (Rayner, 2006). These are human problems that don't often have a right or wrong answer. But these problems can have *better* or *worse* answers, and their study can cumulatively deepen understanding over time, even if the impact is often relatively slow, diffuse and hard won. Different stakeholders are sure they have the answers, but "the trouble is the answers that they have are just irreconcilable with each other" (Rayner, 2006).

Dr. Bill Hooke's blog for the American Meteorological Society, *Living on the Real World*, also considers wicked problems related climate change and environmental hazards. He quotes Rayner as saying that the wicked problems are made even more complex by three factors: 1) time is running out; 2) there is no central authority; and 3) those seeking to solve the problem are causing it (Rayner, 2006). Dr. Hooke agrees with Rayner that, "[t]he truly wicked problems exist not in the inanimate universe of physics, but in the realm of social sciences. Sociologists use this term to describe a class of challenges that societies find themselves poorly equipped to overcome" (Hooke, 2010).

Wicked problems have the following characteristics:

1) the solution depends on how the problem is framed and vice versa so that the problem definition depends on the solution;
2) stakeholders have radically different world views and different frames for understanding the problem;
3) the constraints that the problem is subject to and the resources needed to solve it change over time; and
4) the problem is never solved definitively (Hooke, 2010).

With weather forecasts, the expectation for precision in location and timing of precipitation and storms of all types increases along with the forecast improvements. The rising expectations are accompanied by critical comments that blame the forecasters for over or under forecasting weather—especially severe weather. This is one example of the wicked problems that the study of weather and society faces.

5.6 Hard or Soft Science? Evening the Playground Between the Sciences

It is common to hear social science referred to as a soft science when physical science is considered hard science. This perspective is unfortunately at the heart of many challenges for physical and social scientists understanding each other and working together. Dr. Steve Rayner, an anthropologist at Oxford University, challenges this view stating that physical science is the easy science. Physical scientists can often repeat their experiments, and they have some control over their experimental design, and social scientists cannot readily have the same control in their research (2006).

Ziyad Marar, president of Global Publishing at SAGE Publishing and whose academic background is in philosophy and psychology, uses the argument that social science is more difficult than physical science to call for more and better funded social science rather than less: "The impact of social science may be more diffuse and long term than in much of the natural sciences, but it would be absurd to conclude, with the U.S Senate, that it is a waste of taxpayers' money" (2013). New initiatives from funding agencies including the NSF, IIASA, and many others can foster innovation. Calls for particularly integrated proposals can motivate younger scholars to collaborate in unprecedented ways. Leveraging existing activities, we can increase capacity to make the challenges of integration practical in the near term (Marar, 2013). At the 2011 Weather Ready Nation workshop, Dr. Shirley Laska, a sociologist, bluntly stated that the value of the theoretical frameworks and methods of many social sciences must be accepted before meteorologists and social scientists are partners (University Corporation for Atmospheric Research, 2011).

The conventional wisdom of a simple categorization of science into hard and soft must change. Meteorologists can benefit from qualitative as well as quantitative data. As of 2017, the words "social science" are consistently mentioned in NOAA and National Academy of Sciences reports. The next challenge is to make people aware that social science is systematic, rigorous, and needs to be funded.

5.7 Human Machine Interface

Changing interactions between models, machines, and humans highlight need for integrated approaches. As meteorological forecast models become more technologically sophisticated, the role of the human forecasters has been questioned. As machines and models get smarter, how does human capability to make decisions need to change? The AMS hosted meetings in 2004 related to the automation of forecasts with and without human forecaster inputs.

Many decision makers take specific actions based on specific thresholds, and forecasters provide the specific forecast values. Does the issuance of probabilistic, as opposed to deterministic, forecasts diminish, or enhance the user's perception of the leadership and decision making of the forecaster? "Forecasters will have the most impact by helping design and produce a variety of graphical and text products.... Without the forecasters and management working together in pursuit of the optimum role of the human in the forecast process, the future role of humans will remain very uncertain" (1501).

Since 2011, the NWS has emphasized impacts-based forecasting as part of its Weather Ready Nation initiative. Forecasters are working more closely with local transportation departments, emergency managers, the business community, and others to provide improved forecasts, where the perception of the improvement is in the eyes of the decision makers, not in the eyes of the forecasters or someone who created the forecasting software or model. The emphasis on forecast impacts requires new human-machine collaborations in the NWS.

Dr. Robert Hoffman is a psychologist who works with human factors engineers and others at the Institute for Human and Machine Cognition in Pensacola, Florida. Dr. Hoffman collaborates with weather forecasting software developers to understand how forecasters can maximize the likelihood that new information processing and display systems are user-friendly. The usefulness and usability new technological "decision aids" depend on an understanding of the "knowledge, perceptual skills, reasoning strategies, and decision requirements of the end users" (Hoffman and Coffey, 2004). The human-machine interface professionals are not traditional social scientists or engineers. They have their own methods including "workplace analysis, proficiency scaling, the critical decision method, protocol analysis, knowledge audit and concept mapping" (Hoffman and Coffey, 2004).

In addition to research aimed at improving the research-to-operations relationships, social psychologists and others raise questions about the future of humans as forecasters. Will the models and ensembles be so fast and accurate that the role of the human forecaster will be diminished or eliminated? Will computers replace human forecasters? This topic is often discussed among forecasters and policy makers. NWS's detailed records show a thorough comparison of how well the computers are doing by themselves alongside the value that humans are contributing. In 2012, Nate Silver said, "According to NWS statistics, humans improve the accuracy of precipitation forecasts by about 25 percent over the computer guidance alone. They improve the temperature forecasts by about 10 percent. Humans are good enough, in fact, that when the organization's Cray supercomputer burned down in 1999, their high-temperature forecasts remained remarkably accurate" (Silver, 2012). In what new ways will the human mind and machines directly communicate in a few years? Will there be workable human/machine interfaces where the sensor networks cooperate with each other in seamless ways. How long until the intelligent highway replaces the "unintelligent" highway or that all cars drive themselves? How will weather factor into the adoption and implementation of these new innovations?

Human factors engineering is rich with potential for the new integrated field of weather and science. Sometimes mistaken for social science, it is "[k]nown also as usability engineering, cognitive ergonomics, or user-centered design, human factors is a marriage of psychology and engineering: the application of a scientific body of knowledge about human strengths and weaknesses to the design of technology" (https://en.wikipedia.org/wiki/Human_factors_and_ergonomics). Human factors engineering has been very useful in the development of new forecasting hardware and software for Weather Forecasting Offices. Human factors engineering experiments show ways that new software and hardware can be designed most effectively, accounting for how many images and maps people can keep in mind, how much data they can process at one time, and how quickly people can change from one screen or set of conditions to another (Hoffman *et al.*, 2006). This work is crucial for understanding ways for the forecasters, hardware, and software to work most efficiently as a team to provide the best forecasts possible in the most understandable formats. As artificial intelligence continues to improve, the human-machine interface will require more attention from interdisciplinary teams including social scientists.

Demuth *et al.*'s research (2013) in how people use the NWS webpage changed the graphical symbols for the webpage based on experiments to learn about public preferences and understanding. Their research also showed that the public wanted start and end times added to the forecast content.

The availability of radar images on mobile devices gives millions of people direct access to real-time storm movement. The National Transportation Safety Board meteorologist Paul Suffern notes one drawback is that some private plane pilots use the radar apps on their phones to chart their flight paths without recognizing that the radar image on their app has a time delay, and several accidents have occurred when pilots thought they were intentionally flying around thunderstorms but instead flew into them (Suffern, 2015).

A next step in realizing the usefulness of these and other research findings is for Dr. Hoffman, Dr. Demuth, and others to teach courses in new online formats to bring forecasters, social scientists, stakeholders and others up-to-date with the science, engineering state-of-the-art as well ways to address the serious challenges. A course through Udacity or Coursera would potentially attract many people interested in the machine/human interface of weather forecasting and warnings.

5.8   Questions for Review and Discussion

1 What wicked problem do you consider most pressing? Is it weather related, such as tornadoes or hurricanes? Is weather a component of the problem, as in climate change and coastal populations? Or is it a problem like economic inequality or political instability? Identify the problem as specifically as you can and argue why it's most important and what recommendations you have for beginning to make necessary changes to address the problem.

2 What is the emphasis of your studies? Does your learning cross many disciplinary boundaries or does it focus mainly on one academic discipline? What are your reservations about the proposition that "convergence science" may have positive impacts on changing ways of thinking to reduce loss of life from severe weather because it brings specialists together from different fields? Is it realistic in your lifetime to expect universities and institutions to change in this way? What are some incentives for people to engage in major changes to their existing ways of doing science? How can the convergence science movement contribute to integrated weather and society research and practice?

3 The NWS wants to improve public and official preparations for hurricane and storm surge. What two recommendations do you have for improving their "products?" Do you recommend better maps, better communication, more information, more social science research to find out what vulnerable people already know or don't know about the hazard, some combination of these items or something totally different?

5.9   Using What You've Learned: Homework Assignment From the Chapter

1 Do you have a weather station? If not, check on Weather Underground or Weatherbug websites to see where the nearest stations are to where you live. Check to see if there is local variation between stations within a 10 km radius of your home in terms

of precipitation or temperature or wind. Do you go outside to make your own observations or are you more comfortable looking at gauge and weather station measurements?

2 When the Weather Channel first broadcast "Storm Stories," one of their intended motivations was to use the documentaries or re-enactments of extreme weather impacts to help people identify their own weather vulnerabilities and to encourage viewers to take steps to reduce their own risk whether over the long term by purchasing insurance or buying anchors for their mobile homes. Once viewers observed the disastrous consequences of others' action or inaction, did seeing the events on the screen as an "eyewitness" and seeing interviews with people who had lost homes in floods or tornadoes encourage viewers to be more careful in their own lives? Do you personalize the experiences you see in weather "porn" movies? Or, do the images have a different effect and you don't identify with the subjects in the movies. Do you feel sympathy for the people affected and or empathy for the suffering they are experiencing?

References

Alexander, C. (2013) Storm Shelters in Schools? Wagoner Superintendent Says Safety is Top Priority. KTUL ABC Tulsa. http://ktul.com/archive/storm-shelters-in-schools-wagoner-superintendent-says-safety-is-top-priority (accessed July 23, 2017).

Bainbridge, W.S., and Roco, M.C. (2016) Science and technology convergence: with emphasis for nanotechnology-inspired convergence. *Journal of Nanoparticle Research*, 18: 211.

Barnes, L.R., Gruntfest, E., Hayden, M.H., Schultz, D.M., and Benight, C. (2007) False alarms and close calls: A conceptual model of warning accuracy. *Weather and Forecasting*, 22: 1140–1147.

Brooks, H. (2013) Get as low as you can and put as many walls as you can between you and the tornado. May 30. http://www.livingontherealworld.org/?p=899 (accessed July 23, 2017).

Creutin, J.-D., Borga, M., Gruntfest, E., Lutoff, C., Zoccatelli, D., and Ruin, I. (2013) A space and time framework for analyzing human anticipation of flash floods. *Journal of Hydrology*, 483: 14–24.

Creutin, J.-D., Borga, M., Lutoff, C., Scolobig, A., Ruin, I., and Creton-Cazanave, L. (2009) Catchment dynamics and social response during flash floods: the potential of radar rainfall monitoring for warning procedures. *Meteorological Applications*, 16: 115–125.

Cumbow V. (2010) Some North Alabama schools closing early because of area tornado watch. *The Huntsville Times*. http://blog.al.com/breaking/2010/10/some_north_alabama_schools_clo.html (accessed July 23, 2017).

Demuth, J.L., Morss, R.E., Lazo, J.K., and Hildebrand, D.C. (2013) Improving effectiveness of weather risk communication on the NWS point-and-click web page. *Weather and Forecasting*, 28: 711–726.

Hoffman, R.R., and Coffey, J.W. (2004) Weather forecasting and the principles of complex cognitive systems. *Proceedings of the 48th Meeting of the Human Factors and Ergonomics Society*, 48: 3.

Hoffman, R.R., Coffey, J.W., Ford, K.M. and Novak, J.D. (2006) Notes and correspondence: A method for eliciting, preserving, and sharing the knowledge of forecasters. *Weather and Forecasting*, 21: 416–428.

Holling, C.S. (2001) Understanding the complexity of economic, ecological and social systems. *Ecosystems*, 4: 390–405.

Hooke, W. (2010) Wicked problems. Living on the Real World blog. September 20. http://www.livingontherealworld.org/?p=95 (accessed July 23, 2017).

Layton, L. (2013) Some shelterless Oklahoma schools to cancel class when tornados threaten. July 11. http://stateimpact.npr.org/oklahoma/2013/07/11/some-shelterless-oklahoma-schools-to-cancel-class-when-tornados-threaten/ (accessed July 23, 2017).

Lazo, J.K., Bostrom, A., Morss, R.E. *et al.* (2015) Factors affecting hurricane evacuation intentions. *Risk Analysis*, 35(10): 1837–1857.

Marar, M. (2013) Why does social science have such a hard job explaining itself? http://www.guardian.co.uk/higher-education-network/blog/2013/apr/08/social-science-funding-us-senate (accessed July 23, 2017).

Morrow, B.H., Lazo, J.K., Rhome, J., and Feyen, J. (2015) Improving storm surge risk communication: Stakeholder perspectives. *Bulletin of the American Meteorological Society*, 96: 35–48.

Morrow, B.H. (2013) Personal communication. July.

Morrow, B., and Lazo, J.K. (2013) *Survey of coastal emergency managers perspectives on NWS storm surge information: Hurricane Forecast Improvement Program/Storm Surge Roadmap.* NCAR Technical Note.

Moser, S.C., and Dilling, L. (2007) *Creating a Climate for Change: Communicating Climate Change—Facilitating Social Change.* Cambridge, UK: Cambridge University Press.

National Hurricane Center. (2016) Tropical Storm Hermine. http://www.nhc.noaa.gov/refresh/graphics_at4+shtml/153527.shtml?inundation (accessed July 23, 2017).

Ogle, K. (2013) July 10. http://www.news9.com/story/22799244/oklahoma-school-says-its-buildings-are-tornado-proof (accessed July 23, 2017).

Prevatt, D. (2013) Storm shelter expert shares concerns with Congress. http://today.ttu.edu/2013/06/storm-shelter-expert-shares-concerns-with-congress/ http://www.davidoprevatt.com/?s=uf (accessed July 23, 2017).

Rayner, S. (2006) Jack Beale Memorial Lecture on Global Environment at James Martin Institute for Science and Civilization Wicked Problems: Clumsy solutions – diagnoses and prescriptions for environmental ills. http://www.sbs.ox.ac.uk/research/Documents/Steve%20Rayner/Steve%20Rayner,%20Jack%20Beale%20Lecture%20Wicked%20Problems.pdf (accessed July 23, 2017).

Rittel, H., and Webber, M. (1973) Dilemmas in a general theory of planning. *Policy Sciences*, 4: 155–169.

Roco, M.C., and W Bainbridge, W.C. (eds). (2003) *Converging Technologies for Improving Human Performance: Nanotechnology, Biotechnology, Information Technology, and Cognitive Science.* Dordrecht: Springer.

Roco, M., Whitesides, G., Murday, J., Ferreira, P.M., Ascoli, G., Kong, C.H., Teague, C., Mahajan, R., Rejeski, D., Yablonovitch, E., Cao, J., and Suchman, M. (2013) Methods to improve and expedite convergence. In: Roco, M.C., Bainbridge, W.S., Tonn, B., and G Whitesides G. (eds.), *Convergence of Knowledge, Technology and Society Beyond Convergence of Nano-Bio-Info-Cognitive Technologies*, p. 141.

Ruin, I., Shabou, S., Terti, G., Anquetin, S., Creutin, J.-D., and Lutoff, C. (2014) Flash flooding and the Global Environmental Change perspective: Toward a scaling approach for disaster reduction. Paper presented at the World Weather Open Science Conference, Montreal, Canada. August 16-21.

Ruin, I., Lutoff, C., Creton-Cazanave, L., Anquetin, S., Borga, M., Chardonnel, S., Creutin, J.-D., Gourley, J., Gruntfest, E., Nobert, S., and Thielen, J. (2012) Water and society: A space-time framework for integrated studies. *Bulletin of the American Meteorological Society*, October: ES89–ES91.

Shepherd, M. (2016) 'Turn around don't drown' Is a cute slogan: Why some don't do it. *Forbes Science*, April 19. http://www.forbes. com/sites/marshallshepherd/2016/04/19/turn-around-dont-drown- is-a-cute-slogan-why-some-dont-do-it/#686aac1b3695 (accessed July 23, 2017).

Silver, N. (2012) The weatherman is not a moron. New York Times Magazine, September 7. http://www.nytimes.com/2012/09/09/magazine/ the-weatherman-is-not-a-moron.html?_r=0 (accessed July 23, 2017).

Suffern, P. (2015) The NTSB and forensic meteorology presentation to the Rotary Club of North Raleigh, NC. http://www.slideshare.net/ nraleighrotary/the-ntsb-and-forensic-meteorology (accessed July 23, 2017).

University Corporation for Atmospheric Research (UCAR). (2011) Press release. http://www2.ucar.edu/news/4810/economic-cost- weather-may-total-485-billion-us (accessed July 23, 2017).

Webster, A. (2016) Recognize the value of social science. *Nature*, (7)April: 532.

Wertz, J. (2013) State impact on OETA: Discussing threats to Oklahoma's environment. July 25. https://stateimpact.npr.org/ oklahoma/author/joewertz/page/46/ (accessed July 23, 2017).

Whitesides, G. (2014) Leading scientists discuss converging technologies. http://gmwgroup.harvard.edu/media/index. php?story=120 (accessed July 23, 2017).

Yan, H., and Mier, A. (2016) Houston floods by the numbers. April 16. http://www.cnn.com/2016/04/19/us/houston-flooding-by- the-numbers/index.html (accessed July 23, 2017).

6

Ways to Be Part of the Transformation to Integrated Weather Studies

"If you kick a barge, you hurt your foot. If you lean on a barge, it will start to move."
Dr. William Hooke, WAS * IS workshops (2007–2011)

6.1 Be Part of the Move From WAS to IS

Early career professionals are demanding more balanced training and experience with socially relevant projects, and thus transforming meteorology and hydrology. One striking element that is frustrating and positive at the same time is that academics and practitioners who have been working at the intersection of weather and society for years can get impatient with these newcomers' questions and work. The leaders who are profiled in Chapter 4 are committed to changing the stove-piped uni-disciplinary ways of doing business in meteorology. They know that there will be pushback and challenges from people and institutions that resist change. New bright, energetic people learn about the possibility of integrated studies every day. Each person dedicated to these new ways of doing weather and society business must actively support work that is aimed at the new, integrated weather and society approaches. Those with experience in these approaches must be patience as each person learns and embraces new ideas and ways of thinking at his or her own pace. False starts are a part of the process.

The reference list in this textbook is a first step to a set of essential readings to bring key readings and a common language to all students, faculty members, researchers, software developers, forecasters,

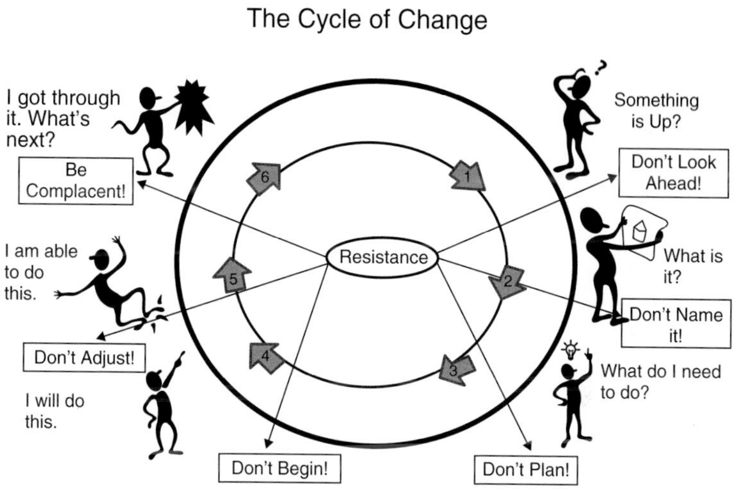

The Cycle of Change

Figure 6.1 Maurer 2009, adapted from Maurer 1996. Reproduced with permission from R. Maurer.

emergency managers, broadcast meteorologists and others engaged in integrated weather–society work, but this is an unrealistic expectation. It is a 2017 snapshot of a rapidly growing field. Expanding social networks, continuing to build and deepen the discussions on the web-based discussion boards, extending the reach of the community beyond common partners at professional meetings including risk communication, sustainability, climate change, and other gatherings will broaden the community in unpredictable ways (Figure 6.1).

6.2   Understand the Cycle of Change

Take a long-term perspective: Don't let the challenges and pushback overwhelm you.

Interdisciplinary work is hard, but it is not impossible. Inventing new ways forward is exhilarating and maddening at the same time. The trajectory and pathways toward a major paradigm are shifting, but years or even decades may be required for an integrated weather/society paradigm to fully emerge. People who decide to move in new directions are often intellectually, institutionally, and geographically isolated in their professional environment at school or work. Resources and time for new work are limited. Understanding new methods and theories and communicating in new ways are both major challenges. This book or a workshop may provide new perspectives and tools to apply to research and practice, a new language, new friends and colleagues, and a motivation and desire to do things differently than before. The movement is growing. Trying to stay optimistic and agile and building new relationships with local, like-minded colleagues are key factors for success. It takes time to learn the languages and perspectives of people from different backgrounds and disciplines. The process will be slow, and progress will be difficult to measure. There will be painful setbacks, including pushback and resistance to change from stove-piped academic and government institutions. The process of personal change parallels changes in disciplinary perspectives (Moser and Dilling, 2007).

The movement to incorporate social science into meteorology is a long and non-linear process, and it does not enjoy unanimous support. As stated earlier, some meteorologists fear that integrating social science theories and methods and even working with social scientists will dilute their "hard" science. Some of the critiques of new approaches will be direct and others will be subtle or contradictory. Some meteorologists say *publicly* that they do support the movement that is integrating social science and meteorology, but behind closed doors, when funding, resources, hiring or other decisions need to be made that would foster the development of the movement, these same professional fall into the traditional ways. Figure 6.2 is a Venn diagram showing perspectives on the integration of meteorology and social science. It simplifies a very complex set of relationships between scientists of many backgrounds. It is not meant to be "insulting" but rather to encourage early career scientists to aspire to tackle integrated projects that involve social and atmospheric sciences.

In 2011, 2012, and 2015, Weather Ready Nation workshops were held (Lindell and Brooks, 2012). Many social scientists were among the invited participants to the meetings held in Norman, Oklahoma and Birmingham, Alabama. Several breakout sessions at the workshop identified several prioritized social science projects. Highly placed administrators and academics gave verbal support for social science. Nevertheless when social scientists present clear evidence in the forms of stories or quotations from emergency managers about how weather information is only part of the suite of information they need to make

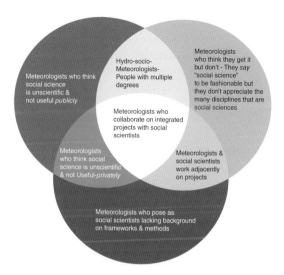

Figure 6.2 Venn diagram showing perspectives on the integration of meteorology and social science.

an informed decision, meteorologists have difficulty "using" this type of information to improve their warning products. Qualitative data is derisively called "anecdotal," and social scientists are frequently asked to quantify their results so they can be directly plugged into the equations of the physical scientists. New work that weaves social science into meteorology and hydrology explores novel methods as well as new questions. The qualitative methods, not often used by physical scientists and engineers, include conducting interviews, conducting surveys, holding focus groups, and participant observation.

Social science theories and methods add important dimensions to the quests to address the challenges at the intersection of meteorology and society. Are meteorologists ready to accept that social science is scientific and that it adds to their fundamental scientific-ness? Some physical scientists (sometimes the brightest ones) think that because they've read about social science methods that they can practice social science based on a few lectures or reading one book. They oversimplify the complex art of writing surveys and are collecting data or conducting interviews without recognition that they would benefit from collaborating with social scientists with experience in the effective uses of these methods.

One reason why there are so few social scientists collaborating with meteorologists is that social scientists *have* credibility as scientists outside of meteorology. When social scientists submit research proposals to programs at the National Science Foundation or the National Institutes of Health, their theories and methods have recognized value. They don't need to be subject to doubt about whether or not their research has theoretical or societal value as they often do when trying to work on weather-related problems. They have their own spheres for research funding, teaching, and application in public policy and practice in emergency management and elsewhere.

6.3 Keep the Momentum Going Toward the Integration of Social and Atmospheric Science

At the WAS * IS workshops, a popular presentation was offered just before the participants left the workshop to return to home, to their regular work, and to their friends and colleagues. Called "the Laundry Talk," it referred to the reality that, while the WAS*ISer was learning exciting new theories and meeting passionate new friends and colleagues, his/her partners and colleagues at home had to handle the everyday tasks, like the laundry, in their absence. When returning home after such a stimulating experience, participants want to exuberantly share their perspectives. However to bring colleagues and family members on board, experts show that the best first thing to do is to thank the people who took care of the mundane chores while they were gone before bombarding them with all the exciting new ideas that were learned.

The "Laundry Talk" also realistically mapped the ups and downs that will accompany the journey to try to get new ideas incorporated into existing institutions or projects. It was meant to give hope and encouragement when the inevitable frustrations and pushbacks arise. The talk helped ease the process of returning to the frustrations of the "real" world, developing ways to make sustainable changes.

The "Laundry Talk" was built upon Kornfeld's 2000 book, *After the Ecstasy, the Laundry: How the Heart Grows Wise on the Spiritual Path* (Kornfeld, 2000). The book addresses how difficult it is to change science, institutions, and other belief systems. Moser and Dilling extended the Kornfeld's perspective to the challenge of communicating climate change. Kornfeld is a well-known Buddhist teacher and someone with positive energy and insights that can be applied to the struggle to change scientific approaches to the intersection of weather and society. Times of great insight and radical new visions for moving forward naturally alternate with life's dirty laundry, "periods of fear, confusion, neurosis, and struggle" (Kornfeld, 2000:xix). The "Laundry Talk" emphasizes that all the changes that need to be made will not be made quickly, that the long journey will have its ups and downs and that it is important to not give up because of pushback. Recognize that not everyone in the office or in the bureaucracy or university is excited about the new work, and it's necessary to deal with resistance or lack of interest. Institutional and scientific changes result from complex processes. Moser and Dilling note two tasks of change agents: 1) elevating motivation to change and 2) lowering resistance/barriers to change. The move forward toward an integration of social science and meteorology requires six key steps: 1) a vision and clear goals; 2) strategic leverage points; 3) windows of opportunity; 4) key players; 5) specific needs to implement the change; and, 6) taking a realistic, but long-term, view (Moser and Dilling, 2007).

6.4 Build Durable Partnerships—Recognize the Power of Networking

The power of networking cannot be overestimated. Try to say "yes" to all opportunities because new surprising directions will emerge. Meetings are time-consuming and major distractions from papers that need to be written or jobs that need to be completed. Often meetings yield unforeseen new collaborators, ideas for approaches or funding, or invitations to other meetings. At a conference, someone might announce an international opportunity, tell stories about students working on innovative research projects, or provide information from agencies that intend to fund some new work.

Sharing new ideas and possible ways to address pressing concerns in open discussions does not characterize most interactions of academics, private sector companies, or even student meteorologists. Such holding back of ideas holds everyone back. Addressing common challenges requires everyone to buy-in on multi-faceted, multi-disciplinary, multi-partner solutions. Early career people need to share what they know and to operate in a transparent manner to build trusting relationships. Asking questions and pushing for clear communication across disciplinary boundaries are worth the considerable effort.

Social science has helped to bring the stakeholder and software developer points of view onto the same platform to improve weather forecasts. Working together maximizes the opportunities for emergency managers, NWS forecasters and others to believe that their ideas will be a valued part of the development of new forecasting tools. They put more time into evaluating materials they receive, and their responses to "what would you like to see in the next version" are more thoughtful than they might have been when they did not feel included as partners in the forecasting software development process.

There is no metric for the professional value of building personal relationships. "Schmoozing," getting to know new people, listening to how they present and interpret their own problems and solutions is essential to moving forward. Not only does the interaction provide refreshing new insights that are applicable, but developing trust with new potential colleagues leads to new and unforeseen collaborations. There's an old adage about how important it is to "show up." This cliché holds true in developing new partnerships between meteorologists and social scientists. Many opportunities will arise from being perceived as an active part of the meteorology and the social science communities.

With a rich network, you can receive valuable and constructive criticism of ideas or written work from professionals whom you trust. Pressures are much greater on students and early career professionals now than they were a generation ago. The pressure can, but does not need to, lead to unhealthy competition. It's difficult not to take criticism personally and to feel not smart enough or a good enough writer for a task. Suggestions from others who you trust make it easier to take critiques in the spirit of improving the content of what is being said and as encouragement for constructing a stronger supporting argument.

The 2011 ethnography (Spinney and Gruntfest, 2012) of emergency managers showed that once the emergency managers knew that their ideas were valued as part of the forecasting software development process, they were more willing to express their ideas. They felt they were a valued part of the process. This step of developing the two-way relationship between the forecasting software developers and the potential users of the new software is very important. Social science has helped to bring the stakeholder and software developer points of view onto the same platform. This allows the emergency managers, NWS forecasters, and others to know that their ideas will be a valued part of the development process. As a result, they put more time into evaluating materials they receive and their responses to "what would you like to see in the next version" are more thoughtful than they might have been when they did not feel included as partners in the forecasting software development process.

6.5 Support May Come From Surprising Sources

From the standpoint of atmospheric scientists it is not possible to tell, in advance, who will be the most important leader or the strongest advocate for work that advances the collaborations between social scientists and atmospheric scientists. Some meteorologists whose words sound most encouraging can lack follow-through and other people who seem doubtful or even hostile to social science have provided all kinds of financial support and other types of support. As the number of people working at the intersection of social and physical sciences increases, new leaders and advocates will emerge from the forecaster, emergency manager, bureaucrat, researcher, student, and other communities. New leadership will bring new outlooks and perspectives. A successful policy entrepreneur can come from any of the partner communities (Roberts and King, 1991). It is imperative for everyone at the intersection of weather and society to be respectful to all partners and be prepared for both disappointments and for happy surprises.

6.6 Five Key Research Priorities for New Hybrid Weather Society Researchers and Practitioners

1) *Broaden the focus beyond "communicating uncertainty" and extending lead time. Recognize that information flows in multiple directions, from forecasters, and also from people in the paths of storms. Recognize that formal weather forecasts are only one component of a complex decision-making process for organizations or individuals*

These issues are part of a larger context. To be sure, taking a narrow view allows the discussions to focus on whether people prefer deterministic or probabilistic forecasts or the difference in decision-maker activities when they have 15 minutes or two hours official warning for a tornado. However, social science research shows that there are instances when the rain/no rain forecast is enough. For many other decisions and in different time and space scales, the probabilistic information is essential. The NWS needs to "communicate what is known."

The 2016 COMET training module is a step in the right direction (COMET, 2016). The training module for forecasters gives examples of the types of interactions between forecasters and emergency managers, transportation managers, and others for various types of weather including snowstorms. It is based on a 2015 YouTube video by psychologist Dr. Susan Joslyn (https://www.youtube.com/watch?v=SfXlt40StpA) to the Workshop on Communicating Uncertainty to Users of Weather Forecasts, held in Whistler, BC, Canada. The module is available from the COMET website.

When forecasters issue warnings, they have assumed that there is "one public" which receives, understands, and responds to the message in the same way. Many studies (e.g., Hayden *et al.*, 2007, Benight *et al.*, 2007) show that there is no one-size-fits-all approach. People hear warnings and respond to them based on their need to know about particular conditions. For example, outdoor field coaches for school matches need to have information to decide whether there may be a lightning risk on the field during the afternoon's high school game on Thursday, but for the rest of the week, a coach would not be as concerned and would not seek such detailed information. Not only is there no unified public, people need different types of weather information during different times of their lives: if they are driving across mountain passes, if they are getting married, or whatever is planned for the day or week. Social science highlights the need for forecasters and emergency managers to consider all possible channels, languages, and ways to inform vulnerable people in time for them to make wise judgments about what behavior is most appropriate to reduce their risk.

With the understanding that people make complex decisions, forecasters will recognize that more than better forecasts are required to reduce death tolls from tornadoes. In 2002, meteorologists Brooks and Doswell showed that the increase in the percentage of the population living in mobile homes in parts of the United States vulnerable to tornadoes meant that there would be an increase in the number of fatalities, regardless of forecasting improvements, because building type matters. From their study of the deaths from the 1999 Oklahoma City tornadoes, they found that mobile home residents were particularly vulnerable. They found that "the rate of death is relatively close to the pre-1925 values in the United States. The increase in use of mobile homes for housing has meant that they are an increasingly large component of the overall death toll" (Brooks and Doswell, 2002:357).

In addition to housing type, other factors can greatly increase vulnerability to these storms, such as: the time of day, day of week, time of year, and/or age. In terms of housing type, other researchers also find that the most vulnerable populations are those living in mobile homes (Simmons and Sutter, 2005; Ashley, 2007; Simmons and Sutter, 2008; Sutter and Erickson, 2010; Chaney and Weaver, 2010). Timing of the storm can increase vulnerability. Several studies have found that nocturnal tornadoes are particularly hazardous, as they are more difficult to identify, the public is less likely to receive a warning because people are sleeping, and residents of vulnerable structures are more likely to be home (Simmons and Sutter, 2005; Ashley, 2007; Black and Ashley, 2011). Tornado fatalities are affected by the day of the week of occurrence, as expected fatalities are about 70% higher on a weekend than a weekday, again presumably because residents of vulnerable housing stock are more likely to be home rather than away at work (Simmons and Sutter, 2008). Research has also illustrated that the month of tornado occurrence has a large impact on the number of fatalities.

Studies by Ashley (2007) and Simmons and Sutter (2008) show that off-season tornadoes may be more deadly because people are more aware of the hazard during the traditional severe weather season (i.e., April to June) and less prepared during other times of the year. Finally, age of victims has also been shown to have an effect on tornado fatalities. People over 40 years old have a higher percentage of fatalities compared to the proportion of U.S. population in that age range (Ashley, 2007).

Local emergency management capability also influences vulnerability. Effective evacuation takes more than notification. People need a safe place where they can evacuate. Someday it will be commonplace for social elements including race, gender, and class to be considered in meteorological-vulnerability research as they are in disaster research (Thomas *et al.*, 2013, Phillips and Morrow, 2007). Vulnerability and resilience are elements outside of NWS control. Major social science research efforts that examine the context and not just the NWS products plus more collaboration between forecasters, media, and local emergency managers will also help reduce vulnerability and improve the likelihood that people will change their behavior in timely and appropriate ways.

After the deadly 2011 U.S. tornadoes, Dr. Jeff Lazo said: "I would wager that most if not all of these events were reasonably well forecasted and warned for. I would also wager that feasible near-term improvements in forecast skill for most of these events would not have significantly reduced the social or economic impacts and would not have appreciably saved lives or reduced injuries. I would further wager that a much greater return on investment-measured in terms of lives saved and damage avoided-could come from research, applications, and operations to improve the communication, understanding and response to forecast and warning information than could come from research to improve forecast skill, false alarm rates, or even lead time" (Lazo, 2012; http://www.sip. ucar.edu/news/volume6/number1/director_note.php).

Too much energy is spent on the question of whether to provide deterministic *or* probabilistic information. Since there is no one-size-fits-all weather consumer, both kinds of information are essential. Current debates that focus on "the communication of uncertainty" feed into an unbalanced dialogue. Individuals deal with probabilistic information on a daily basis. Decisions about when to take a particular action or particular route if traveling all require complex considerations of uncertainty. Are tourists willing to take protective behavior when weather warnings are issued (Jeuring and Becken, 2013)? What people consider acceptable summer weather in Scandinavia is the focus of a study published by Denstadli *et al.* in 2011. Weather is not a major barrier to tourism in high latitude regions (Jeuring and Becken, 2013). They found that most people visiting Scandinavia in the summer do not adjust their plans based on weather forecasts. Tourists to northerly areas have realistic images of destination weather. Characterizations of "good" and "bad" weather diverge between various types of visitors. Most visitors follow their original plans in spite of various weather conditions.

Weather-related questions that people ask are, for example, will there be less traffic if I take a smaller, longer route rather than taking the major highway at rush hour? Should students be held at school for an extra 15 minutes to wait until the heavy rain from the thunderstorm lets up? These questions posed millions of times everyday require more than just "better," "more accurate," or more "precise" weather information.

Some social science research is underway to evaluate what probabilistic information of particular-sized and -shaped polygons for warnings or ways of conveying uncertainty are most valuable. There is an official but misguided quest for one particular answer about which kind of information, format, or source will be best. Forecasters and operations researchers ask: "Is a rectangle more meaningful than a pointy polygon? Is this yellow and green combination communicating best or should other colors be used?" However, this quest is misguided since the social science evidence that does exist indicates that people use a wide range of information and their specific needs vary during the course of their lifetime or their daily responsibilities.

Millions of dollars are being spent to increase forecast lead time for tornadoes. Where is the evidence that additional lead time will have benefits for people? Research findings indicate that providing longer forecasting lead time may or may not be appreciated by emergency managers or the public. Longer lead times do offer emergency managers, hospital administrators, school administrators, highway departments, and others the chance to stage equipment or change daily operations in preparation for severe weather. How will weather-sensitive decision makers, including emergency managers and school administrators, use that time and geographic specificity to change their daily operations when severe weather is expected? How might this finer resolution, spatial, and temporal data be adaptable to specific users needs? How might longer lead times reduce the vulnerability to severe weather of specific populations and individuals? Will increased lead time result in different decision making, such as a change in the location people choose to go or whom people are contacting for confirmation?

Numerous new partners from different intellectual traditions as well as from multiple agencies are needed to solve contemporary problems at the intersection of weather and society. More work that considers the process from end-to-end-to-end (Morss *et al.*, 2005) is necessary. The traditional, top-down perspective from research to operations should be replaced by a platform with all partners having an equal voice from the research to the applications and from applications to research. Improved warnings and response will require active collaborations between forecasters, local emergency managers, school superintendents, road departments, and others. The collaborations will change the common equations for partnerships to be more inclusive, beyond traditional disciplinary or agency stovepipes. Conceptual models are being challenged and should lead to new policies in many aspects of meteorological research and practice.

Traditional ways of verification within the NWS undervalue the services that forecasters provide. Forecasters are now evaluated based on a hit or miss system that does not account for events that occur as forecasted but are not as severe as expected or are in a slightly different location than the forecasted location. New research shows that most people understand that there is uncertainty in each forecast (Morss *et al.*, 2008). With new communication capabilities and wide internet usage, *too much* information, including conflicting information, poses a problem in some instances just as a lack of information does about pending hurricanes, tsunamis, volcanic eruptions, flash floods, or other hazards.

How do homeowners decide whether to update their kitchens with granite countertops or to improve their roof to survive a major tornado? Dr. David Prevatt, a wind engineering professor at the University of Florida, studies wind damage to residential buildings. He studied the damage after the 2011 Tuscaloosa, Alabama and the Joplin, Missouri tornadoes. In his testimony before the House Science, Space and Technology Subcommittee on Research on June 5, 2013, he recognized that reducing a home's vulnerability to a tornado does cost money, but he doesn't buy the argument that the costs are beyond homeowners' means. He pointed out that homeowners find money to update their kitchens with granite countertops but don't find the same $10,000 to improve their home's resilience to tornadoes (http://today.ttu.edu/2013/06/storm-shelter-expert-shares-concerns-with-congress/).

2) *Change the research and media focus from why did so many people die to how did the many thousands of people* survive *the severe weather events*

This second research priority for new hybrid weather society researchers and practitioners is a fairly radical departure from studying what failed to studying what succeeded. In 2011, when more than 600 people died in tornadoes in the United States, it was easy for politicians and the media to blame bad forecasts for the large death toll. In spite of excellent lead times and accuracy, the media can always find someone who tells the camera that "The tornadoes came without warning." Not everyone hears or pays attention to official warnings, but all the 2011 storms were forecast with state-of-the-art precision.

When the death toll is studied alongside study of the magnitude and intensity of the tornadoes, a better question would be, "How did so many people survive these incredible storms?" What actions did *survivors* take? What motivated them to take those actions? The old research conversations usually began with the technical warning times and boundaries. New research must focus on the context in which people heard and responded to warnings and to environmental cues. Many people do receive and respond to official warnings. Look at how many lives warnings have saved in addition to paying attention to the actions of the people who are killed. No matter how much money is spent on improving forecasts, as has been noted earlier in this book, there are many factors besides warnings that affect survival rates. It is impossible to reduce the death tolls to zero.

More than 32,000 people were killed in 2014 in car accidents (National Highway Traffic Safety Administration, 2015). Car accident deaths can be reduced significantly if people drove three miles per hour and car bumpers were many times larger than they are now, but the tradeoffs of making these time or design compromises are not acceptable. Similarly, it is not possible to reduce weather-related deaths to zero, but it is possible to significantly reduce the number of people who are injured or who die. Learning more about what motivates many people to heed warnings and move out of harm's way will be a constructive social science research direction.

3) *Change from the passion for advancing the next technological innovation to a passion for doing work that addresses expressed stakeholder needs*

Instead of asking why didn't everyone respond to formal warnings in timely and effective ways, turn the question around. Ask what types of information are weather-sensitive decision makers using and wanting, and what motivates them to change their behavior when facing severe weather threats? Build research programs that start with the needs of weather-sensitive decision makers at all scales, rather than on the new technological products. Create new research questions, new frameworks and models, and build upon what already has been developed. More research is needed on how all types of decision makers use weather forecasts now and on how can they better use them. League, Hoekstra, Spinney, and Nichols have all completed research projects recently that shed light on what temporal and spatial scales emergency managers at universities, cities, and public schools prefer. Starting with the expressed stakeholder and the public preferences for what people want and use will be more productive than continuing to develop more products without input from the intended consumers of the new products. When people feel that their opinion matters, they will be more helpful and provide more complete answers to questions about what weather information and its timing is working or not working. The pleasantries of superficial interactions need to change to a more meaningful dialogue.

Until recently, computer programmers working at forecasting software laboratories created new forecasting software and changes in forecasting operations at NWS offices were driven by new technological capabilities. Forecasters and others would be forced to use new tools that were designed, in the eyes of the programmers or in some cases bureaucrats, to improve their ability to do their jobs. When forecasters resisted the new tools as being too cumbersome, complex, or even less functional than what they already were using, new training was arranged. Forecasters would be "trained" to use the new tools. If new software more directly addressed forecaster and stakeholder concerns, new software would be more readily accepted and adopted more quickly and forecasts might be more accurate and timely. The Hazardous Weather Testbed and the FACETS program at the National Weather Center in Norman are addressing this issue by bringing stakeholders directly into their work earlier in the software development process than they did a decade ago.

4) *Research what motivates people to change their behaviors and decision making*

Recent behavioral science studies (e.g., Ruin, 2008; League, 2009) show that weather information is only one piece of a puzzle that influences how drivers behave when faced with heavy rain that might flood roads along their route. Some people are more weather-sensitive than others. Many factors influence how weather sensitive a person is on any particular day. With regard to driving across flooded roads, many people have reported that they know there are warnings in effect and they know that other people have died at particular low water crossings but they say they "have" to go—whether it's to get to work or to pick up their children. In these cases, it is more than a matter of "better information" to encourage people to change their route or the time of their trip. Another example of a new method is to move beyond work that studies perceptions to work that studies actual behaviors (e.g., Ruin *et al.*, 2014; Benight *et al.*, 2007; Hayden *et al.*, 2007). So much of the weather/social science asks people "what would they do if…"? With the recent proliferation of public cameras and YouTube videos research can focus on what people actually do rather than on what they say they would do.

Predicting human behavior is possibly more difficult than forecasting future atmospheric conditions. As more meteorologists and hydrologists learn that social science methods are scientific and are quite similar to what *they* practice in terms of theoretical frameworks, observations, and analysis, more productive partnerships will grow. Practicing social science requires a firm background in methods and theory. Writing a survey or conducting focus group meetings requires a set of skills and experience that are common to the social sciences. It is more realistic for meteorologists to partner with social scientists than to expect that they will be able to "be" social scientists. Emphases on perception research should be replaced with emphases on behavior research. What people say they would do may not reflect what they actually do. New research should start from the behaviors rather than from the environmental threat. This is a radical suggestion, particularly when researching the physical threats receives most of the research funding. While social science, looking at behaviors after events, has increased in scope and in number of case studies, the resources are weighted toward the physical science end. What if major research efforts started with the actions of vulnerable segments of the population or with the actions taken by emergency managers or transportation departments?

Dr. Ruin and her colleagues have developed new set of research guidelines allowing interdisciplinary research in the post-event context (Ruin *et al.*, 2014). When a flood occurs and there are teams set up to learn from the experience, there is a focused effort for all team members, regardless of their background, to learn a diverse set of data from their field work. For example, traditionally hydrologists would enter the field in a post-flood assessment and would investigate hydrological impacts including the rate of flow and the height of the floodwaters. Without the new guidelines, any information the hydrologists received from the field, say from a business owner about what actions people took, rather than the actions of the river, would not be considered extraneous data and would not be recorded. Such data would be invaluable for social scientists wondering why people did what they did and when they took particular actions. With the new guidelines, the hydrologists would keep track of stories about what people did and what decisions they made during the flood, just as the social scientists in the field record any data they learn about the height or speed of the floodwaters when they do post-flood fieldwork. This commitment to a multi-disciplinary approach for studying floods has tremendous potential for understanding floods, not just the hydrologic or the meteorological aspect but also in more integrated and comprehensive ways (Creutin *et al.*, 2009; Creutin *et al.*, 2013; Ruin *et al.*, 2014; Ruin *et al.*, 2012). Figure 6.3 provides examples of the types of questions that some of the various social science disciplines can ask individuals to build social science and meteorology collaborations. These are a small subset of the types of questions related to agencies, social groups, and events that can be asked.

5) *Improve the warning process with comprehensive and inclusive research efforts*

Many research projects can address how warnings can best reach the highest proportion of people vulnerable to a particular type of severe weather. Is there an over warning or under warning problem—and if so, for whom? How can warnings be improved? What about the false alarm "problem?" Is there such a thing as warning fatigue, and if there is, can forecasters and others minimize its impact? Mackie's 2014 dissertation on warnings for bushfires in Australia shows how optimism bias and warning fatigue hamper official efforts to effectively warn communities at risk (Mackie,

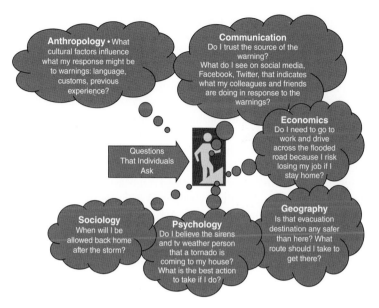

Figure 6.3 Questions social scientists from many disciplines might ask about individual human behavior in severe weather.

2014). Is there a continuum for warnings rather than the NWS's hit or miss options (Barnes *et al.*, 2007)? New research based on many focus groups and interviews aims to identify better ways of messaging and communicating (Eosco, 2015). New media, including Twitter, Facebook, and others keep all the partners and the public up to the minute with real-time reports and photos, and the processes and opportunities for communication are expanding as social media changes. How can we best use these media to protect vulnerable populations?

6.7 How Science Changes

Social science has actively been incorporated into physical science research in socio-hydrology. This successful integration moves beyond the side-by-side approach of research and can be a role model for full integration of social sciences and weather fields.

6.8   Socio-Hydrology Emerges

The weather community, including atmospheric and social scientists, is one of many scientific groups that are trying to be more integrative of new partners and new disciplines. Atmospheric scientists and hydrologists consider similar problems, but often even in the NWS, their problems are not approached in an integrated fashion. The meteorologists worry about the atmosphere and the precipitation, but they do always learn what happens when that rain or snow hits the ground and vice versa.

Recent scholarly publications define and expand fields called "socio-ecology" and "socio-hydrology" (Sivapalan *et al.*, 2012) and socio-meteorology (BASC Committee on Progress and Priorities of U.S Weather Research and Research-to-Operations Activities, 2010:42). These studies show that to comprehensively understand flooding, hydrological and social scientific knowledge are essential. There is a long way to go before meteorology and hydrology are fully integrated. However, there has been notable formal progress in bringing social science and hydrology together. Sivapalan and his colleagues are the pioneers of socio-hydrology (Sivapalan, 2009). According to Sivapalan, there are more than 80-refereed publications in the new field and 20 Ph.D. students around the world who are specializing in socio-hydrology for their doctoral theses (Sivapalan personal communication, 2016).

One example of socio-hydrology is the study of the Murray Darling watershed in Southeast Australia. They studied hydrologic, agricultural, and political changes in the water resource demand and availability in the basin. "Principals of integrated water resources management, an established field, allowed planners to consider past, present and future irrigation needs for agriculture and environmental protection purposes. While integrated water resource management is about the interactions of humans and water and often uses the "scenario-based" approach," Sivapalan *et al.* (2012) point out that the integrated water resource management approach is not socio-hydrology. Traditional work does not "account for the dynamics of the interactions between water and people" (Saveniji and Van der Zaag, 2008:1271).

"Socio-hydrology" is defined as the science of people and water that is "aimed at understanding the dynamics and co-evolution of coupled human-water system." (Milly *et al.*, 2008; Peel and Bloschl, 2011) In socio-hydrology, humans and their actions are considered 'part and parcel' of water cycle dynamics, and the aim is to predict the dynamics of both" (Sivapalan *et al.*, 2012:1271).

The socio-hydrologists also engage in the new science of eco-hydrology. Eco-hydrology explores the co-evolution and self-organization of vegetation in the landscape in relation to water availability. "Socio-hydrology... explores the co-evolution and self-organization of people in the landscape, also with respect to water availability" (2012:1271). Sivapalan *et al.* believe that socio-hydrology will "take on increasing importance in the context of a changing, human-dominated world, its practice may turn out to be more challenging than eco-hydrology ...because humans possess more powerful ways and means of controlling water cycle dynamics beyond the optimality, adaptation and acclimation strategies that natural vegetation possesses and has developed over time"(2012:1272). Sivapalan and his colleagues conclude that socio-hydrology will change how hydrologists do their scientific work. They will expand their collaborations beyond their existing partnerships to build a more holistic understanding of hydrologic science (2012:1275).

6.9   New Integrated Disciplines on the Horizon

New integrated frameworks will allow new scientific approaches and theories to emerge that address problems at the intersection of weather and society. The new disciplines will be different than meteorology, hydrology, or any of the social sciences. A transformation to new scientific disciplines would be similar to the emergence of the fields of material science or nanotechnology. They are well accepted but they began as a result of numerous collaborations and changes in research priorities.

The NOAA strategic plan calls for the inclusion of social sciences, and there is a cadre that includes hundreds of ambitious early career meteorologists and hybrid meteorologists who are committed to changing the stove-piped culture in sustained ways. Historians of science believe that most changes in science are incremental rather than revolutionary and that the integration of social science into meteorology may fit into that model (Kuhn, 1962). Metrics of the move toward integrated socio-meteorology will be useful to show the tempo and the nature of the changes that are occurring.

This timeline shown as Figure 6.4 highlights new promising initiatives, including WAS * IS and SSWIM that have been defunded. As of 2017,

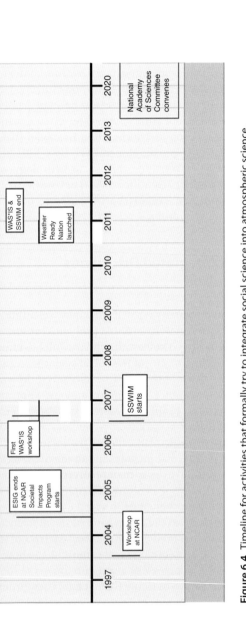

Figure 6.4 Timeline for activities that formally try to integrate social science into atmospheric science.

there is growing evidence that the weather enterprise is moving forward toward integrated physical/social science directions. Many individuals, universities, companies, agencies and others are trying new projects that evaluate social media and other new communication methods for reaching diverse publics (Demuth *et al.*, 2013, Eosco, 2015; Klockow, 2013).

A commitment to socio-meteorology as a discipline similar to socio-hydrology may be useful. For the reasons specified by the socio-hydrologists, an even more comprehensive approach of socio-hydro-eco-meteorology may be the most useful approach to solve complex, dynamic contemporary and future weather-related problems.

6.10 Changing the Paradigm Takes Time and Will Require Patience

Many of the research studies already underway or recently completed in 2017 are first steps. NCAR's research scientists including Drs. Morss, Demuth, Lazrus, Hayden, and Wilhelmi are accelerating progress with their large, innovative work. All of their recent findings should be considered as baseline data for future case studies. Building on what already has been done is key to moving forward (Lazrus *et al.*, 2016) and the work of many pivotal thinkers is cited the reference list.

Workshops and curricula based on an integrated perspective and highlighting successful case studies will help create new social scientists to collaborate on weather-related problems in a truly integrated fashion. Just as atmospheric scientists need to know something about and appreciate social science methods, social scientists need to understand more of the theoretical frameworks used in meteorology (and hydrology), the methods, the technologies, and the institutional contexts. They will also have to learn many new acronyms to be fully engaged.

Meteorologists have decades of relatively well-funded research and technology development that have brought their side of the enterprise to its current advanced state of knowledge about how the atmosphere behaves. Understanding how people behave when faced with weather-sensitive decisions and developing effective means to provide them the information they say they need cannot happen quickly. More case studies and other research completed by many teams will move the enterprise forward toward meeting its goal of reducing losses of lives and property.

There are no official metrics to track the progress of the integration of social science and atmospheric science. Anecdotal evidence highlights some of the key concepts and issues that reflect the movement's dedication to big issues and to fundamental culture change to integrate social science. Regardless of the budget problems, pushbacks, and setbacks, the ambitious early career people who are engaged in integrated physical and social science work will accelerate the trend and provide new, imaginative directions and initiatives. Growing demand for participation in grass roots efforts, including WAS * IS (Weather and Society * Integrated Studies) is transforming meteorology and hydrology as early career professionals demand more balanced training and experience with socially relevant projects (Demuth *et al.*, 2007). As of September 2017, there are 1000 people on the WAS*IS Facebook page exchanging ideas, publications, job opportunities, and making new connections. In 2016, a new WAS*IS Students Facebook page was set up mostly for and by students who entered the profession after the 2011 formal end of the WAS*IS program. These students want to learn about and practice weather and society work whether or not WAS*IS is actively funded.

Social science and social scientists are being woven into the fabric of atmospheric science. Culture change is underway. A wide range of decision makers, product developers, faculty, researchers, forecasters, and others are engaged in the development of new warning tools, in new national and international collaborations across the private and public sectors, in new social media communication, and in new decision-maker–driven definitions of severe weather and false alarms.

6.11   Hybrids By Design—Socio-Hydro-Meteorology

This book has narrowly focused on the intersection between meteorology and the social sciences. This is an artificial or constructed focus considering how the meteorological, hydrological, and social systems are interconnected on the planet. Starting with meteorology and the social sciences is a first step of bringing many examples together in one book. One important next step is to broaden the discussion to include hydrology, ecology, health, and other disciplines. The NWS has meteorological and hydrological forecasting responsibilities. Some of the WAS * ISers and others engaged in work already bring hydrologic considerations into the discussions. Dr. Russ Schumacher's 2013-2014 Studies of Precipitation, flooding, and Rainfall Extremes Across Disciplines, or SPREAD, workshop developed an active network of eager, early-career meteorologists, engineers, economists, hydrologists, sociologists, historians, and others to confront this problem directly. SPREAD's goals were to foster innovative thinking beyond disciplinary borders about hydrometeorological, engineering, and societal aspects of extreme precipitation and flooding; to promote mutual understanding of multidisciplinary research and operational tools and methods; to initiate multidisciplinary research collaborations among students with different disciplinary training; and to develop a community of researchers and practitioners working to better understand, predict, and respond to floods.

SPREAD brought attention beyond meteorology to address many weather–society challenges with hydrologic factors as the main consideration. The 2013 and 2014 SPREAD workshops brought together 20 motivated graduate students who were working on research related to some aspect of extreme precipitation, flooding, floodplain management, engineering, societal, and behavioral aspects of precipitation or floods. These graduate students have an active interest in multidisciplinary approaches and perspectives.

Dr. Schumacher called his workshops SPREAD because "when forecasting rainfall, flooding, elections, or anything else, there is always uncertainty. One way this uncertainty is represented is by the "spread" (i.e., the diversity of different possible outcomes) in an ensemble of forecasts. The workshops considered ways to understand and communicate that spread or uncertainty in different contexts (http://schumacher.atmos.colostate.edu/spreadworkshop/index.php). The SPREAD workshop participants have published a collaborative article based on their interdisciplinary work that lays out a flash flood intensity index similar to the scales used for hurricane and tornado (Schroeder *et al.*, 2016).

As of 2017, most of the people dedicated to integrating social science into meteorology and hydrology are still early in their careers. In the next few years and decades, creative integrated new ways will be developed for weaving social science into meteorology, hydrology, and climate change. The new "hybrids" will reinvent the work of academic and international, federal, state, regional, and local agencies so that the stovepipes and older rigid ideas and conceptual models that characterized the last century will no longer be the norm.

6.12   How to Become a Participant in the Movement to Integrate the Social Sciences and Atmospheric Science

As of 2017, there is no undergraduate or graduate program that provides a skill set that includes social and physical aspects of atmospheric science. One option that some people are selecting involves getting one degree in meteorology and a bachelor's or master's degree in anthropology, geography, or communication. A combination of two degrees gives a graduate a bona fide multi-disciplinary perspective. There are also examples where people major in meteorology with a minor in sociology or another social science degree.

One option is to find a graduate program where it is possible to find an advisor who will encourage cross-disciplinary collaborations. As of 2017, it is considered risky for graduate students to choose to focus their research at the intersection of weather and society. It is easier to get a Ph.D. and get a job with a focused narrow disciplinary perspective and become an expert within atmospheric science, meteorology, or a social science. However, some people are being successful in accomplishing an interdisciplinary degree within traditional programs with the help of a supportive advisor. There is an imaginable future where a student who graduates with a Ph.D. in meteorology goes before a graduate committee with a narrow focus that advances one element of a model or only considers the physical characteristics of a particular storm or weather phenomena will face scrutiny and criticism for NOT taking into account the societal relevance, the "so-what" elements of their atmospheric science research.

The 2015 American Meteorological Society definition of a Bachelor's Degree in meteorology, which includes a heavy course load of math and physics requirements, allows little room for a broadening the students' education. Steps toward this change include developing a broader university curriculum for undergraduate- and graduate-level seminar classes that introduces the integration of meteorology and social science and developing online courses to teach various integrated meteorology-hydrology-social science topics. New integrated graduate programs may emerge soon. Early career scientists who want to earn advanced degrees that complement their bachelor's degrees or who want to find a program with integrated weather/climate social/physical science in its approach have no good options for graduate programs.

Many eager students piece together a program in geography. The intense interest in the WAS * IS program and email inquiries show there is growing demand by bright physical scientists who recognize how complementary training in social science or communication can enhance their expertise as a forecaster or researcher. Developing new programs will take years and will be met with resistance for budgetary and philosophical reasons. Dr. Amber Silver earned her Ph.D. in geography at the University of Waterloo in Canada. Stephanie Hoekstra earned her Ph.D. in coastal resource management at East Carolina University (Hoekstra, 2015).

6.13 The Challenges Are Not Overwhelming

6.13.1 Scientific Budgets Are Lean

In 2017, scientific funding for all types of research and operations is tight. Many programs are being cut and others are cobbling together funds to stay afloat. Because these are tight budgetary times, the pushback against social science funding may be part of a larger context where physical scientists as well as social scientists are competing for reduced funds. Research and operations budgets are under pressure. New programs are unlikely to be embraced when agencies and funders have to make sure existing projects, personnel, and infrastructure, are maintained. Hopefully progress will draw from with the formal recognition and support of social science in agencies' Strategic Plans, the growing cadre of ambitious early career professionals with meteo-socio-hydro backgrounds, the new research landscape that includes physical and social scientists at the table, and recognition that new integrated verification and evaluation metrics are needed.

6.13.2 Be Patient and Optimistic and Keep Pushing

The first steps toward collaboration take a lot of time. New workable language needs to be found or developed since disciplines and agencies all have their own languages and ways of thinking. Contemporary efforts to change meteorology to be more inclusive of social science must build on earlier programs. Social scientists must have incentives to work with meteorologists. If meteorologists do not recognize anthropologists, sociologists, and others, as "scientists" with equal theoretical and methodological footing, then social scientists will continue to work independently on weather risk, weather communication, and other problems that interest them. To encourage both sides to work together will require constant pressure from scholars excited by the results and possibilities of these collaborations.

6.14 Meteorologists Embracing Social Science Is a First Step

There is often a sense of euphoria when meteorologists first consider how social science can help them better understand how people hear and respond to their forecasts. This euphoria can be followed by a letdown and a sense of frustration when they realize, upon some reflection and some research, that the social sciences are so diverse, that there is little communication across social science disciplines, and that integrating social science is not as simple as finding social science collaborators interested in weather problems. Another sense of frustration comes from the recognition that identifying how social science can provide essential answers to difficult questions about societal impacts of weather still requires extensive, expensive, and time-consuming social science research to address questions related to how to increase the number of people who take appropriate and timely actions when faced with potentially severe weather.

The 40 years of progress in the fields shows us the middle ground between euphoria and frustration. The vocabulary of all the partners in the weather enterprise is more inclusive now. The NWS has formally included societal impacts as integral parts of its forecasting mission. There are dozens of early career professionals dedicated to the end-to-end process that includes social and physical scientists and practitioners. There is a growing appreciation, at the highest levels of research and government, that the research-to-operations model benefits from including stakeholders as new software and forecasting tools are developed.

The American Meteorological Society began publishing a journal with an emphasis on societal impacts of weather in 2008: *Weather, Climate and Society*. WAS * ISers, COMET, and others are developing new hydro-meteo-socio-courses on line and at universities (i.e., Sherman-Morris at Mississippi State University, Ng at University of Colorado at Denver, Godfrey at University of North Carolina Asheville, and Stevermer at COMET).

There is a growing social science presence at the annual American Meteorological Society (AMS) meetings. In 2016, the eleventh annual Symposium on Policy and Socio-Economic Research Policy sessions was held in New Orleans, Louisiana. Attendance at most of the sessions has risen to standing room only. Including social sciences as part of the student sessions of the AMS has had an enormous impact of cultivating the interest of hundreds of enthusiastic meteorology students in the social sciences. Leaders from WAS * IS, water managers, private sector weather consultants, and many others have been invited presenters and exposed the students to a wide variety of career options in government, academia, non-profit sectors, and the private sector. Topics have included water, transportation, energy, natural hazards, and many others. For many meteorology students at the AMS meeting who are locked in very narrow curricula in their home departments, the AMS sessions are their first inkling of how to apply their meteorological expertise in the real world.

6.14.1 Be Part of the Integrated Studies Movement

One striking element that is frustrating and positive at the same time is that academics and practitioners who have been working at the intersection of weather and society for years can get impatient with newcomers' questions and work. Consider the leaders who are profiled in Chapter 4. They are committed to changing the stove-piped, uni-disciplinary ways of doing business in meteorology. While some university officials embrace integrated approaches, there will be pushback and challenges from people and institutions that resist change. New people learn about the possibility of integrated studies every day. Each person dedicated to these new ways of doing weather and society business must actively support work that is aimed at the new integrated weather and society approaches, even if it is done by a complete beginner.

6.14.2 Work Out Wide-Open Ways for Partners to Thrive

The private weather sector is growing. Numerous discussions are underway that are trying to draw formal lines of responsibility for the public and the private sector. Based on the research that shows a wide variability in stakeholder weather information needs, there is plenty of work to be done by all sectors. The public deserves excellent forecasting and scientific support services that they already pay for with their taxes.

The private sector has an endless set of potential clients ranging from transportation companies, cities, sports stadiums, retail stores, and many more. As personal mobile devices become more pervasive, building new mobile applications will provide work for all partners. Many of the contemporary efforts require close collaboration between the private and public sectors. One example is Weathercall (http:// weathercallservices.com/) a service that lets you know when the NWS has issued a tornado warning or severe thunderstorm warning for your location. Many of the discussions on Facebook, Twitter, at Integrated Warning Team meetings, in the press, and elsewhere are heated and complex.

Some meteorologists think the enterprise would be better off if all or most the work was transferred to the private sector rather than keeping vibrant public and private sectors. Some see the federal government's role only as a data provider rather than one with a close interaction with public decision makers. If this work falls entirely on the private sector, we will lose a key element in these productive new partnerships. What is difficult to measure but is significant is the trusted, respectful and comfortable ways the teams collaborate. Teams can be found at the local, regional, national, and global scales.

6.15 Work for the Longer Term

6.15.1 New Integrated Graduate Programs and New Integrated Training Models

Physical scientists and social scientists are schooled at perfecting and preferring one perspective: one way of seeing the world. Atmospheric scientists, just like any traditionally trained uni-disciplinary scholar, favor their own ways of asking and answering questions. As long as these stovepipes continue to be the mainstream ways of thinking and the main ways of training at universities and professional schools, the prospects for integrated perspectives are limited. There are powerful institutions including Arizona State University and the National Science Foundation that are trying to change the uni-disciplinary models. Unfortunately, most people still consider themselves as uni-disciplinary specialists, rather than scientists dedicated to work at the intersection of weather and society. Until new interdisciplinary perspectives are developed, with incentives built in for career advancement and adopted, major reductions in the impacts of severe weather will be difficult to accomplish.

One key for developing a career in this field is to participate in professional meetings. Consider interdisciplinary professional meetings such as the Association of State Floodplain Managers (www.floods. org), the Association of American Geographers (www.aag.org), the Association of State Dam Safety Officials, the ALERT User Group, or the National Hydrologic Warning Council (www.hydrologicwarning. org). Professional organizations usually have student registration rates, and many universities will support student participation in professional conferences. Going to student and regular meetings of professional associations including the American Meteorological Society (www.ametsoc.org) is productive. To get a broader perspective of the new work at the intersection of social science, meteorology and hydrology consider the European Geophysical Union (www.egu.eu) or the HyMex annual meeting (www.hymex.org).

Many WAS*ISers report that their interest in becoming a professional meteorologist was based on an intense weather experience they had when they were younger. For example, their family's farm was affected by drought, a hurricane or severe snowstorm caused severe disruptions in daily life, a tornado roared through their town, or the extreme cold or heat affected them personally. What meteorological events can you credit with helping to shape your interest in this field, especially related to the intersection of weather and society?

6.15.2 Positive Outlook for Careers for People with a Background in Integrated Studies

With backgrounds in meteorology and in aspects of social science, graduates are finding employment in agencies and companies engaged in natural hazard loss reduction. "Demand is spiking for scientists and engineers in the multidisciplinary field of natural hazards risk analysis. Experts in the field, which focuses on how to predict, prevent and limit damage from disasters, are finding jobs in insurance, agriculture, finance, infrastructure, construction, humanitarian aid and public policy." "Good students…are having no trouble finding positions in those areas. Particularly fruitful areas of study include earth sciences, physical sciences and sustainable development" (Nelson quoting Arthur Lerner-Lam, the deputy director of Columbia University's Lamont-Doherty Earth Observatory in Palisades, New York, 2013).

Opportunities in risk analysis, another growing interdisciplinary field, fall mainly into three overlapping categories: the natural and physical sciences, engineering, and social sciences. Engineers assess infrastructure and sometimes inform policy decisions; social scientists often study how best to communicate risk. Natural and physical scientists, from geologists and meteorologists to mathematicians and physicists, study the origins, movement, and potential impacts of natural hazards. Some reinsurance companies, which house insurance to insurance companies, offer employees in-house training related to the job requirements. For example, the giant reinsurance company, Swiss Re, trains employees in financial modeling. The U.S. Army Corps of Engineers hires people formally trained as hydrologists, as geomorphologists, and as social scientists, and then they also offer cross-disciplinary training to employees (Nelson, 2013:273).

Calls to bring social science and atmospheric science to the same page are not new. In 1971, more than 45 years ago, findings from a National Academy of Sciences report said: "The committee recognizes that thorough assessments of the benefits of weather forecasting or of other weather services do not exist, that dissemination of information to the public is largely based on traditional procedures using outmoded technology and that there is often insufficient interaction between the user and the information" (National Research Council, 1971:2). There will be a great deal of work for many people for years to come.

6.15.3 Climate Change will Exacerbate the Need for Integrated Weather–Society Research and Practice

In 2016, the four-year California drought is only one of many examples of how the past weather records may not be reliable sources for basing future weather and climate predictions. Even the major 2015-2016 El Niño that was nicknamed the "Godzilla El Niño" because of intensity failed to live up to the high expectations. It brought only "average" rainfall amounts to the northern part of California. "Average is no drought-buster" (Famiglietti, 2016). NASA satellite data show that it will take at least two or three more years of average or above average rainfall for formally end the current drought (Famiglietti, 2016). Jay Famiglietti is senior water scientist at the Jet Propulsion Laboratory who wrote the op-ed piece for the *Los Angeles*

Times that ends with the following quote that is apt for California and elsewhere and eloquently makes the case for the importance of expertise at the intersection of social science and atmospheric science: "The California drought will end, but it is a preview of a drier future here and beyond our borders. Population growth, changing climate and disappearing groundwater have converged to a point where large swaths of the country and the world are facing permanent water losses. California can lead by example. But we must act quickly to pursue the social, financial, technological and governance innovations that chronic water scarcity demands." This image from the NOAA Drought Monitor shows the severity of the western United States drought conditions even after the winter El Niño precipitation. During the 2016-2017 winter, rains eased the California drought conditions.

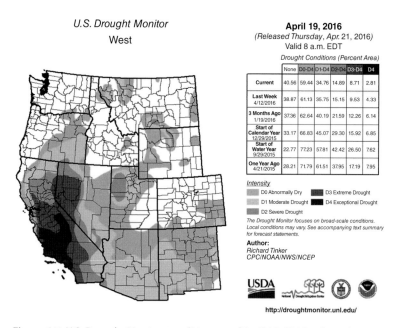

Figure 6.5 U.S. Drought Monitor conditions as of April 19, 2016, released April 21, 2016 (http://droughtmonitor.unl.edu/home/regionaldroughtmonitor. aspx?west). Showing how much of the western United States remains in extreme drought. *Source:* NDMC-UNL, 2016. Reproduced with permission from The Drought Monitor, National Drought Mitigation Center, University of Nebraska-Lincoln.

6.16   Questions for Review and Discussion

1   Imagine you are the mayor of the city that will be hosting the summer Olympics. What are the weather conditions you worry about the most in terms of hosting the games effectively? What particular weather sensitive decisions do you need to make? What kinds of data would help you most with your weather decision making? Anderson-Berry *et al.* (2004) summarized the trials and tribulations of forecasting the 2002 Sydney, Australia games. They found that the organizers were just as concerned with when the adverse weather would end as they were with when it would begin. For example, would the rain and wind calm down enough to allow the kayak competition before the Olympic closing ceremonies? What factors, other than the weather, are extremely important to you in your daily decision making? How will the demand for weather information change in the near and the long term?

2   How weather sensitive are you? How often do you check official forecasts? What are your favorite sources for weather forecasts? What characteristics make these sources your favorites? Accuracy, timeliness, ease of access, or mobile availability? What other factors? Does your active interest in the weather depend on the time of the year or what particular activities are on your schedule—days of your vacation? What about particular sporting activities?

3   What specific suggestions do you have for changing the culture of meteorology to be more accepting of the integration of social science? Do you think it is important to make these changes to effectively reduce losses from severe weather? How would you allocate funds if you were the director of a major scientific funding agency?

6.17   Using What You've Learned: Homework Assignment From the Chapter

1   In Australia, New Zealand, most of Europe and elsewhere, there is not as much distinction between weather and climate issues as there is in the United States. This book steers away from climate change issues with its focus primarily on weather. Dr. Marshall Shepherd uses analogies to help people understand the difference:

"People tell me all the time that there's no such thing as global warming because it's cold outside," he said. "So I tell them to think of it this way: weather is like your mood and climate is like your personality. Your mood shifts from day to day, just like our weather. But climate is the big picture; it's what's happening in the long term" (Shepherd quote in Hataway, 2015). Dr. Shepherd also uses a baseball analogy: "Think of bad storms like home runs in baseball's recent steroid era. Sure, the big hitters have always hit home runs, and I can't say that any one homer resulted from steroid use. However, I can make the argument that steroids made a lot more balls go over the fence than normal." "Climate change is no different. You have to have multiple data points to make a solid argument, and I think that we will probably see more violent storms because of rising global temperatures. But it's disingenuous to suggest that one storm is the result of climate change" (Hataway, 2015). See more at http://www.uga.edu/about_uga/profile/uga-scientist-refutes-climate-change-skeptics/#sthash.sGZemfiK.dpuf. Do you think there should be separation between consideration of how to cope with severe weather and adaptation to climate change or do you believe the distinction is artificial? As far as policy change recommendations for science, does it make more sense to consider climate change and weather as one entity? Support your answer.

2   Look up on the web the NOAA drought monitor map for today. How have the conditions changed since the April 2016 example shown above? (Figure 6.5). In what places has the drought intensified and where has it weakened? You can see current drought monitor maps at http://droughtmonitor.unl.edu/.

3   When you think of your career, how do you think of yourself? Are you someone who thinks that you can make the biggest impact by staying within the boundaries of your discipline or your particular job, or do you imagine yourself as someone who will try to break down walls and barriers between groups and ways of thinking? There are benefits and tradeoffs with each option. Do you think there is a happy middle ground where you can push for change but also maintain the status quo? Do you believe that maintaining the status quo at least assures that the movement toward the integration of social science and atmospheric science does not go backward to even more rigid siloed ways of thinking?

References

Anderson-Berry, L., Keenan, T., Bally, J., Pielke, R., Jr., Leigh, R., and King, D. (2004) The societal, social, and economic impacts of the world weather research programme. Sydney 2000 forecast demonstration project (WWRP S2000 FDP). *Weather and Forecasting*, 19: 168–178.

Ashley, W.S. (2007) Spatial and temporal analysis of tornado fatalities in the United States: 1880-2005. *Weather and Forecasting*, 22: 1214–1228.

Barnes, L.R., Gruntfest, E., Hayden, M.H., Schultz, D.M., and Benight, C. (2007) False alarms and close calls: A conceptual model of warning accuracy. *Weather and Forecasting*, 22: 1140–1147.

Benight, C., Gruntfest, E., Hayden, M., and Barnes, L. (2007) Trauma and short fuse weather perceptions. *Environmental Hazards*, 7: 220–226.

Black, A.W., and Ashley, W.S. (2011) The relationship between tornadic and nontornadic convective wind fatalities and warnings. *Weather, Climate, and Society*, 3: 31–47.

Board on Atmospheric Science and Climate (BASC), Committee on Progress and Priorities of U.S Weather Research and Research-to-Operations Activities. (2010) *When Weather Matters: Science and Service to Meet Critical Societal Needs*. National Research Council. http://www.nap.edu/catalog.php?record_id=12888 (accessed July 23, 2017).

Brooks, H., and Doswell, C. (2002) Deaths in the 3 May 1999 Oklahoma City tornado from a historical perspective. *Weather and Forecasting*, 17: 354–361.

Chaney, P.L., and Weaver, G.S. (2010) The vulnerability of mobile home residents in tornado disasters: The 2008 Super Tuesday tornado in Macon County, Tennessee. *Weather, Climate, and Society*, 2: 190–199.

COMET. (2016) Communicating Forecast Uncertainty. https://www.meted.ucar.edu/training_module.php?id=1225#.VxfLD5MrIdU (accessed July 23, 2017).

Demuth, J.L., Morss, R.E., Lazo, J.K., and Hildebrand, D.C. (2013) Improving effectiveness of weather risk communication on the NWS point-and-click web page. *Weather and Forecasting*, 28: 711–726.

Denstadli, J.M., Jacobsen, J.Kr.S., and Lohmann, M. (2011) Tourist perceptions of summer weather in Scandinavia. *Annals of Tourism Research*, 38(3): 920–940.

Eosco, G. (2015) Town Hall Meeting: Watch out! A review of the National Weather Service's watch, warning, advisory hazard messaging system. It's advised you attend. You have been warned! American Meteorological Society meeting, Phoenix, AZ, January 5.

Famiglietti, J. (2016) Is the California drought America's water wake-up call? *Los Angeles Times*, April 16.

Hataway, J. (2015) Zombie slayer. *UGAresearch*, Spring: 21–26.

Hayden, M., Drobot, S., Gruntfest, E., Benight, C., Radil, S., and Barnes, L. (2007) Information sources for flash flood warnings in Denver, CO and Austin, TX. *Environmental Hazards*, 7: 211–219.

Hoekstra, S.H. (2015) *Decisions under duress: Influences on official decision making during Superstorm Sandy.* Ph.D. thesis, Coastal Resource Management, East Carolina University.

Hoekstra, S.H. (2012) *How K-12 school district officials made decisions during 2011 National Weather Service tornado warnings.* Master's thesis, Department of Geography and Environmental Sustainability, University of Oklahoma.

Jeuring, J., and Becken, S. (2013) Tourists and severe weather – An exploration of the role of 'Locus of Responsibility' in protective behaviour decisions. *Tourism Management*, 37: 193–202.

Klockow, K. (2013) *Spatializing tornado warning lead-time: Risk perception and response in a spatio-temporal framework.* Ph.D. thesis, Geography, University of Oklahoma.

Kornfeld, J. (2000) *After the Ecstasy, the Laundry: How the Heart Grows Wise on the Spiritual Path.* New York: Bantam Books.

Kuhn, T.S. (1962) *The Structure of Scientific Revolutions.* Chicago: University of Chicago Press.

Lazo, J. (2012) NCAR Societal Impacts Program. http://www.sip.ucar.edu/news/volume6/number1/director_note.php (accessed July 23, 2017).

Lazrus, H., Morss, R.E., Demuth, J.L. *et al.*, (2016) Know what to do if you encounter a flash flood: Mental models analysis for improving flash flood risk communication and public decision making. *Risk Analysis*, 36(2): 411–427.

League, C.E., Philips, B., Bass, E.J., and Diaz, W. (2012) Tornado warning communication and emergency manager decision-making. Presentation at American Meteorological Society, January 24. http://ams.confex.com/ams/92Annual/flvgateway.cgi/id/20175?recordingid=20175 (accessed July 23, 2017).

League, C., Díaz, W., Philips, B., Bass, E.J., Kloesel, K., Gruntfest, E., and Gessner, A. (2010) Emergency manager decision making and tornado warning communication. *Meteorological Applications*, 17: 163–172.

League, C. (2009) *What Were They Thinking? Using YouTube to Observe Driver Behavior While Crossing Flooded Roads.* Unpublished master's thesis, Applied Geography, University of Colorado at Colorado Springs.

Lindell, M.K., and Brooks, H. (eds.). (2012) Workshop on Nation: Science Imperatives for Severe Thunderstorm Research Report, held April 24-26 in Birmingham, Alabama.

Mackie, B. (2014) *Warning Fatigue: Risk Communication and the Australian Bushfires.* Ph.D. thesis, University of Canterbury.

Milly, P.C.D., Betancourt, J., Falkenmark, M., Hirsch, R.M., Kundzewicz, Z.W., Lettenmaier, D.P. *et al.* (2008) Stationarity is dead: Whither water management? *Science*, 319: 573–574.

Morss, R.E., Wilhelmi, O.V., Downton, M., and Gruntfest, E. (2005) Flood risk, uncertainty, and scientific information for decision-making: Lessons from an interdisciplinary project. *Bulletin of the American Meteorological Society*, 86: 1593–1601.

Moser, S.C., and Dilling, L. (2007) *Creating a Climate for Change: Communicating Climate Change- Facilitating Social Change* Cambridge University Press

National Highway Traffic Safety Administration. (2016) 189 NCSA data resource website. Fatality analysis reporting system. http://www.nsc.org/NewsDocuments/2017/12-month-estimates.pdf (accessed July 23, 2017).

National Research Council Committee on Atmospheric Sciences. (1971) *The Atmospheric Sciences and Man's Needs Priorities for the Future.* National Academies, Washington DC, p. 2.

Nelson, B. (2013) A calculated risk - scientists and engineers with an analytical bent are sought-after in natural-hazard risk assessment. *Nature*, 495: 271–273.

Nichols, A.C. (2012) *How university administrators made decisions during National Weather Service tornado warnings in the spring of 2011.* Master's thesis, Department of Geography and Environmental Sustainability, University of Oklahoma.

Peel, M.C., and Bloschl, G. (2011) Hydrological modeling in a changing world. *Progress in Physical Geography*, 35(2): 249–261.

Phillips, B. and Morrow, B. (2007) Social science research needs: Focus on vulnerable populations, forecasting, and warnings. *Natural Hazards Review*, 8: 61–66.

Roberts, N.C., and King, P.J. (1991) Policy entrepreneurs: Their activity structure and function in the policy process. *Journal of Public Administration Research and Theory*, 2: 147–175.

Ruin, I., Creutin, J.-D., Anquetin, S., and C Lutoff, C. (2008) Human exposure to flash floods – Relation between flood parameters and human vulnerability during a storm of September 2002 in Southern France. *Journal of Hydrology*, 361: 199–213.

Savenije, H.H.G., and Van der Zaag, P. (2008) Integrated water resources management: Concepts and issues. *Physics and Chemistry of the Earth*, 33(5): 290–297.

Schroeder, A.J., Gourley, J.J., Hardy, J., Henderson, J., Parhi, P., Rahmani, V., Reed, K., Schumacher, R.S., Smith, B.K., and Taraldsen, M.J. (2016) The development of a flash flood severity index. *Journal of Hydrology*, April 8 (online).

Simmons, K.M., and Sutter, D. (2005) WSR-88D radar, tornado warnings, and tornado casualties. *Weather and Forecasting*, 20: 301–310.

Simmons, K.M., and Sutter, D. (2008) Tornado warnings, lead times, and tornado casualties: An empirical investigation. *Weather and Forecasting*, 23: 246–258.

Sivapalan, M. (2016) Personal communication.

Sivapalan, M., Hubert, H., Savenije, G., and Blöschl, G. (2012) Socio-hydrology: A new science of people and water. *Hydrological Processes*, 26: 1270–1276.

Sivapalan, M. (2009) The secret to 'doing better hydrological science': Change the question. *Hydrological Processes*, 23: 1391–1396.

Spinney, J. and Gruntfest, E. (2012) *What makes our partners tick? Using ethnography to inform the Global System Division's development of the Integrated Hazards Information Services (IHIS)*. Report prepared for NOAA Integrated Hazards Information Systems Project http://www.evegruntfest.com/SSWIM/pdfs/Final-rep-1.pdf (accessed July 23, 2017).

Spinney, J.A., and Pennesi, K.E. (2012) When the river started underneath the land: social constructions of a 'severe' weather event in Pangnirtung, Nunavut. *Canada Polar Record*, 1–11.

Sutter, D., and Erickson, S. (2010) The time cost of tornado warnings and the savings with storm-based warnings. *Weather, Climate, and Society*, 2: 103–112.

Thomas, D.S.K., Phillips, B.D., Lovekamp, W.E., and Fothergill, A. (2013) *Social Vulnerability to Disasters, 2nd Edition*. New York: CRC Press.

Appendix A

List of Acronyms

AMS	American Meteorological Society
BASC	Board on Atmospheric Science and Climate, National Research Council
CASA	Collaborative Adaptive Sensing of the Atmosphere
COMET	Cooperative Program for Operational Meteorology, Education and Training, University Corporation for Atmospheric Research Boulder, CO
CNH	Coupled Natural and Human Systems program National Science Foundation
CSU	Colorado State University, Fort Collins, CO
DSS	Decision Support Services
DoD	Department of Defense
EM	Emergency Manager
ERG	Eastern Research Group
FAR	False Alarm Rate
FEMA	Federal Emergency Management Agency
GIS	Geographical Information System(s)
GDP	Gross Domestic Product
HyMeX	Hydrologic Cycle in Mediterranean Experiment
IIASA	International Institute for Applied Systems Analysis
IDSS	Integrated Decision Support Services
IHIS	Integrated Hazard Information Services
LTHE	The Laboratoire d'étude des Transferts en Hydrologie et Environnement in Grenoble, France
MSC	Meteorological Service of Canada
NASA	National Aeronautics and Space Administration
NCAR	National Center for Atmospheric Research, Boulder, CO

NIH	National Institutes of Health
NOAA	National Oceanic and Atmospheric Administration
NRC	National Research Council
NSF	National Science Foundation
NWS	National Weather Service
PH.D	Doctor of Philosophy Degree
SEES	Science, Engineering and Education for Sustainability program
SIP	Societal Impacts Program
SOO	Scientific Operations Officer
SPC	Storm Prediction Center
SSWIM	Social Science Woven into Meteorology
USDA	United States Department of Agriculture
USGS	United States Geological Survey
WAS*IS	Weather and Society Integrated Studies
WMO	World Meteorological Organization

Appendix B

Blogs and Websites that Integrate Weather and Society

http://www.washingtonpost.com/blogs/capital-weather-gang
Capitalweathergang: Jason Samenow, Dan Stillman, and their colleagues at the *Washington Post* provide the insightful and entertaining weather coverage for the metropolitan Washington, DC region. Their forecasters and writers tell you what's going on with local weather and why. They explain what can be expected for specific upcoming events and locations of interest. They discuss what causes various weather phenomena, warn about bad (and good) weather on the horizon, and they keep their blog up to date during severe weather or emergencies. The blog is interactive. Readers can join the conversation by posting comments, telling storm stories, asking questions or reporting observations in real-time. The site features:

1) Outlooks for school closures and travel disruptions
2) Weekend and event forecasts, including for football and baseball games
3) Seasonal outlooks
4) Post-mortem forecast assessments
5) Discussion of local, national and international weather news
6) Commentary on climate change and environmental science
7) Questions and answers with local and national weather personalities
8) Forecast contests
9) Humor and off-topic posts, and,
10) Original weather photography.

Weather and Society: Toward Integrated Approaches, First Edition. Eve Gruntfest.
© 2018 John Wiley & Sons Ltd. Published 2018 by John Wiley & Sons Ltd.

http://www.livingontherealworld.org
Dr. William H. Hooke has been a senior policy fellow at the American Meteorological Society since June 2000, and associate executive director of the AMS Policy Program since July 2001.

http://weatherbrains.com/weatherbrains
James Spann, Bill Murray and their colleagues in Alabama lead what they call the Official Netcast For People Who Love Weather. *WeatherBrains* is a weekly audio show delivered by the Internet. The show unites weather geeks worldwide. They do their best to cover the world of weather in a fun way with great guests. The show runs 60 to 80 minutes every week. It is recorded on Monday nights. The show is usually available on the web by midnight Monday nights EST. As of April 2016, they have over 400 shows available.

http://www.wunderground.com/blog/
Weather Underground directed by Dr. Jeff Masters and Robert Henson links to thousands of weather stations and many other essential weather related comments and observations.

http://www.climatecentral.org/blogs
Dr. Heidi Cullen and her colleagues at Climate Central have web resources and blogs related to extreme weather with emphasis on the link to climate change, energy, climate science, and where resilience and sustainability meet. They use the latest web-based and video-based communication techniques and engage social science in integrated ways. From climatecentral.org: "The climate crisis isn't just some far-off threat: it's a clear and present danger. Galvanized by this sobering reality, Climate Central has created a unique form of public outreach, informed by our own original research, targeted to local markets, and designed to make Americans feel the power of what's really happening to the climate. Our goal is not just to inform people, but to inspire them to support the actions needed to keep the crises from getting worse." They are active tweeters and keep viewers up-to-date with the latest media and scientific information related to climate change. Follow Cullen on Twitter (@heidicullen).

http://eloquentscience.com/category/blog
Dr. David M. Schultz is a reader at the Centre for Atmospheric Science, School of Earth, Atmospheric, and Environmental Sciences, The University of Manchester. He is Chief Editor for *Monthly Weather Review*, cofounder and Assistant Editor for the *Electronic Journal of*

Severe Storms Meteorology, Associate Editor for *Atmospheric Science Letters,* and on the Editorial Board of *Geophysica.* He has published over 90 articles. He is the author of *Eloquent Science: A Practical Guide to Becoming a Better Writer, Speaker, and Atmospheric Scientist* published by the American Meteorological Society in 2009. His blog includes numerous helpful hints.

http://www.ccrfcd.org/publicinformation.htm/
Clark County (Nevada) Regional Flood Control District's site is an excellent starting point for state of the art public awareness flash flood campaigns—see billboard competition example.

https://www.atxfloods.com/
Flash flood warning and low water crossing site for Austin, Texas maintained by the City of Austin Flood Early Warning System (FEWS) team.

Appendix C

Other Relevant Web Resources

American Meteorological Society
www.ametsoc.org and the American Meteorological Society Board on
Societal Impacts https://www.ametsoc.org/stac/index.cfm/boards/
board-on-societal-impacts/board-of-societal-impacts-image-gallery/

CASA Collaborative Adaptive Sensing of the Atmosphere
http://www.casa.umass.edu/
Cities on Volcanoes conference http://www.citiesonvolcanoes9.com/

Community Collaborative Rain, Hail and Snow Network (CoCoRahs)
www.cocorahs.org

National Weather Association Committee on the Societal Impacts of
Weather and Climate of the National Weather Association
http://www.nwas.org/committees/societalimpacts/

Weather, Climate and Health group at NCAR
https://www.ral.ucar.edu/csap

Weathercall (a service that lets you know when the NWS has issued a
tornado or severe thunderstorm warning for your location)
weathercallservices.com

Weather and Society: Toward Integrated Approaches, First Edition. Eve Gruntfest.
© 2018 John Wiley & Sons Ltd. Published 2018 by John Wiley & Sons Ltd.

Addendum

Physical and social scientists are recognizing there are no "natural disasters." Professor of Atmospheric Science, Dr. Kerry Emanuel, wrote for the *Washington Post* on September 19, 2017: "The term 'natural disaster' is a sham we hide behind to avoid our own culpability. Hurricanes, floods, earthquakes and wildfires are part of nature, and the natural world has long ago adapted to them. Disasters occur when we move to risky places and build inadequate infrastructure."

In Dr. Ilan Kelman's words on August 30, 2017 in the online *Vox Populi* following Hurricane Harvey: "A disaster involving a hurricane cannot happen unless people, infrastructure, and communities are vulnerable to it. People become vulnerable if they lack knowledge, wisdom, capabilities, social connections, support, or finances to deal with a standard environmental event such as a hurricane. This can happen if lobbyists block tougher building codes, planning regulations, or enforcement procedures. Or if families can't afford insurance or the cost of alternative accommodation if they evacuate. Or if limited hurricane experience induces a sense of apathy."

As this book goes to press at the end of 2017 the impacts of weather are unprecedented. In August, South Asia floods killed 1,200 people and kept 1.8 million children out of school. In the United States, damages from Hurricanes Harvey, Irma, and Maria exceed hundreds of billions of dollars in Texas, Florida, the Southeastern United States, the Virgin Islands, and Puerto Rico.

When you read this you may remember that on September 25, 2017, five days after the Hurricane Maria storm, more than 3 million Americans in Puerto Rico are living in "apocalyptic" conditions ever since the well-forecast storm devastated the entire island. Most people have no electricity or running water. Communication on the island

and with the mainland is challenging and many, perhaps millions, have lost their livelihoods. Also, today, Chicago is experiencing its sixth day of record temperatures above 90 degrees accompanied by dangerous air quality readings. Summer extreme temperatures extend into autumn across much of the nation including Portland, Oregon, San Francisco, California, and Washington, DC.

In all of these rare cases the meteorological forecasts have been excellent for the storms and for the heat. The NOAA atmospheric scientists, forecasting teams, private sector meteorologists, as well as media from TV to Twitter courageously made and conveyed forecasts for the recent extreme events. Hurricane Harvey forecasts for over 60 inches of rain were verified, exceeding amounts that had been recorded before. Accurate forecast tracks for Hurricanes Irma and Maria, when they were in the Caribbean threatening millions of people on the islands and mainland, were combined with findings from social science research to help federal and city officials with their evacuation decision making, and millions of people took successful protective actions that saved their lives.

As a result of this excellent work, why can't WAS*ISers and their colleagues step up and show policymakers and funders how the application of recent research results limited the recent storm death tolls to the dozens rather than in the thousands as they were during the 2005 Hurricane Katrina in New Orleans, Louisiana, and along the Mississippi coastline in ways that bring increased funding opportunities? Our community has not been successful at taking credit for our accomplishments and research scientists and practitioners are busy doing their jobs rather than lobbying for funding.

Will the unprecedented weather lead to a renewed dedication to research that changes the culture of the work at the intersection of weather and society? Possibly.

However, the weather is not the only aspect of our challenging work that is unprecedented. Disdain for scientific expertise and hostility toward governmental agencies fundamentally limits scientists' capabilities to fulfill their missions to weather forecasting, environmental protection, healthy air quality, and other public goods that are within the scientific state of the art in the social and physical realms. Scientific methods and theoretical frameworks based on facts are being challenged in the public and private sectors. These capabilities are being undermined as many positions are unfilled and agency leadership undercuts recent progress on many fronts. With the current political context, agencies and researchers cannot take the bold stances that can change the culture to embrace new collaborations between atmospheric and social scientists.

Rather than building on the successes to date or improving ways to connect vulnerable populations with information they want and can use, funds for much of the most important research and practice at the intersection of weather and society that can build genuine collaboration between the social sciences and atmospheric sciences are being significantly cut and undermined. There are new players emerging, and some efforts are alive but are continuing in weakened conditions at the National Center for Atmospheric Research and elsewhere.

This manuscript began as an effort to chart progress in a blossoming new context of socio-meteorology. While there are bright spots and reasons for cautious optimism, after decades of effort, a "paradigm shift" or integration of social and atmospheric sciences is more remote than it has been in at least a decade.

Two hopes (dimming from an expectation, but nonetheless existing because of the bright, passionate atmospheric and social scientists) are 1) that the demand for integrated efforts from emergency managers, flood plain managers, and other stakeholders will drive new large research efforts at the institutions now doing the work and at other universities and research institutes and 2) that the agencies and institutes who allocate resources to the mission of integrating social and atmospheric scientists will build teams that include new hybrid scientists and will not continue to unrealistically expect major changes with only one person devoted to the work per institution.

The starting point for a possible second edition of this textbook should be a full constructive synthesis and honest appraisal of ways that social scientists and atmospheric scientists are engaged or can be engaged in addressing the largest issues of reducing vulnerability to weather events. Perhaps the contents and conversations will be placed directly in the context of environmental justice, or planning resilient communities, or adaptation to climate change or in some starting point that has not yet been identified or defined? Perhaps the new collaborations and descriptions will be built on terminology that has not yet been specified, understood or accepted. I'm hoping that the "three steps up, two steps back" progression of the work described in this book will change in ways that bring atmospheric scientists and social scientists together on a new collaborative platform and that the integration of social sciences and atmospheric sciences will firm up rather than continue to limp along or collapse.

Index

references to figures are in *italics*

a

adaptation, to climate change *67*, 120
adjacent projects 134–139
airlines, use of meteorology 75–76
American Meteorological Society (AMS)
  Board on Societal Impacts 5, 36
  career progression 185
  conferences 9
  social sciences 182, 183
  Summer Policy Colloquium 39
  web resources 146, 199
anthropology *60*, *62*, *63*, 64–68, *173*
  leader profiles 119–120, 128–129
Arctic, changing conditions 65, *67*, 68
artificial intelligence 150
atmospheric science. *see* meteorology
Australia, Murray Darling watershed 174
automated forecasts 148–149

b

BASC (Board on Atmospheric Science and Climate) 14–15, 111, 174
Baton Rouge floods 2
Becker, Julia *108*, 108–109
behavior 71, *81*, 87
  in crisis 78, 169–170, *173*
  difficulty of influencing 47
  predicting 171
  response to severe weather 4, 6, *173*, 177
behavioral phenomena 60, 62
behavioral science 170–171
big data 143
blizzard warnings 3, 116
blogs 37, 82, 117, 135–136, 146, 196–198
Board on Atmospheric Science and Climate (BASC) 14–15, 111, 174
Board on Societal Impacts, AMS 5, 14, 36, 182, 199
British East India Company 29
British Royal Society 29
building trust 134–135

Weather and Society: Toward Integrated Approaches, First Edition. Eve Gruntfest.
© 2018 John Wiley & Sons Ltd. Published 2018 by John Wiley & Sons Ltd.